Gerald E. Aardsma, Ph.D.

©2017 by Gerald E. Aardsma

All rights reserved. No part of this publication may be reproduced, stored in a retrieval system or transmitted in any form or by any means: electronic, electrostatic, magnetic tape, mechanical, photocopying, recording or otherwise, without prior written permission of the publisher.

Published by:
Aardsma Research & Publishing
412 N Mulberry St
Loda, Illinois 60948-9651

www.BiblicalChronologist.org

Scripture quotations are taken from the New American Standard Bible® (NASB), copyright©1972 by The Lockman Foundation, www.Lockman.org, except as otherwise noted.

Printed in the United States of America

Library of Congress Control Number: 2017911953

ISBN 978-0-9647665-8-7

Contents

List of Tables   5
List of Figures   7
Dedication   9
Acknowledgments   11
Preface   13
1 Beginnings   17
2 The Data   27
3 The Biblical Life Expectancy Graph   35
4 The Central Hypothesis   41
5 Properties of Vitamin X   49
6 The Environmental Abundance of Vitamin X   55
7 The Natural Synthesis and Distribution of Vitamin X   63
8 What the Flood Broke   67
9 Vitamin X Revealed   71
10 Modeling the Biblical Life Expectancy Data   79
11 Of Mice and Men   91
12 Potential Longevity   97
13 Growing Youthful   103
14 Dose Rate   107
15 Personal Testimonial   113
Appendix MeP_20170105.F95   119
Index   133

LIST OF TABLES

1   Selected biblical life span data.   29
2   Birthdates of selected biblical males.   31
3   Point estimates of life expectancies from Adam to Moses.   33

LIST OF FIGURES

1	Biblical life expectancy at birth data for selected males.	34
2	Measured temperature versus time for a bowl of hot water.	37
3	Structure of the vitamin C molecule.	43
4	The Eber–Peleg Drop.	48
5	Noah's life span relative to pre-Flood average.	54
6	Sudden increase and drop in abundance of vitamin X.	57
7	The post-Peleg decline and the Moses Drop.	58
8	Pre-Flood, Spike, Decline, and Modern time periods.	61
9	Oceanic phosphate and nitrate surface distribution.	70
10	The MePA and MeP molecules.	77
11	Computer model results.	81
12	Rate of "aging" versus MePA body pool size.	85
13	Pre-Flood, break-even, and saturation body pool sizes.	87
14	The good fit of the model to the data.	89
15	Survivorship curves of fruit flies.	93
16	Survivorship curves of mice.	93
17	Jump up in life expectancy 4444 B.C.	96
18	Survivorship curve for U.S. males.	98
19	Survivorship curve once "aging" has been removed.	99
20	Healing of agedness.	102
21	Noah's physiological agedness versus calendar year.	105

Dedication

Helen,

How many times have we looked at one another these past seventeen years and recited those words from David Mamet's film, *The Winslow Boy*, "Are we crazy, you and I?"

Thank you for choosing to walk this difficult road with me. You have been a friend encouraging me to press on, a secretary running the business, a comrade covering my back, a nurse watching by my bedside, a business partner financing her research scientist, a teammate setting me up for the shot, ... and all the while a poverty-line housewife looking after her children and her man. How can I not admire you?

Well, now we have the answer. It seems we were not crazy after all. And I am so looking forward to growing young with you.

Gerald

Acknowledgments

I wish to express my sincere appreciation to the following individuals who helped with the task of bringing this book to publication.

My wife, Helen, proofread the first draft. The second draft was proofread and commented on by other family members: my brother Allen and his wife Robin Aardsma, my daughter Jennifer and son-in-law Steve Hall, my daughter Rebekah and son-in-law Mark David Eschbach, my daughter Rachel and son-in-law Joey Contreras, my son Matthew and daughter-in-law Esther Aardsma, and my two sons at home, Timothy and Caleb. Rebekah Eschbach assisted with the creation of the index. Steve Hall provided graphic design and artwork, including the cover design. My long-time friend, Tom Godfrey, provided invaluable editing assistance once again.

Preface

> *So at last Faramir and Éowyn and Meriadoc were laid in beds in the Houses of Healing; and there they were tended well. For though all lore was in these latter days fallen from its fullness of old, the leechcraft of Gondor was still wise, and skilled in the healing of wound and of hurt, and all such sickness as east of the Sea mortal men were subject to. Save old age only. For that they had found no cure...*[1]

I am a research scientist, not a communicator, and most certainly not a salesman.

I have spent most of my life researching at the interface of science and the Bible. My science specialty is physical dating methods such as radiocarbon. My earliest full-time Bible/science research effort centered around the question of why nobody had ever been able to pin a functional historical date on Noah's Flood. This led, eventually, to the discovery that exactly 1000 years had been accidentally dropped from traditional biblical chronology due to an inadvertent copy error in a number found in 1 Kings 6:1.[2]

This rapidly led to answers to other Bible/science questions I had not even set out to answer. For most of the final decade of the twentieth

---

[1] J. R. R. Tolkien, *The Lord of the Rings* (Boston: Houghton Mifflin Company, 1987), *The Return of the King*, 136.

[2] Gerald E. Aardsma, *A New Approach to the Chronology of Biblical History from Abraham to Samuel*, 2nd ed. (Loda, IL: Aardsma Research and Publishing, 1993). www.BiblicalChronologist.org.

century, I was immersed in research connected to the Exodus of Israel from Egypt, the ensuing Conquest of Canaan, and the much earlier in time Flood of Noah. I found that the Exodus was a real historical event, and that the biblical description of it was simply historical.[3] The scholars were all saying the opposite, but that was because they had their biblical chronology wrong by 1000 years. You cannot find an object if you spend all your time looking a thousand miles from where it is located, and you cannot find a historical event if you spend all your time looking 1000 years from when it happened.

I found that the Flood, too, was a real historical event, and that the biblical record of it, too, was simply historical.[4] In fact, I found that the Genesis record of the Flood had to be the result of an eyewitness knowledge of that event.

Concurrent with all of this research, I pursued a lifelong interest in why, according to Genesis, humans had lived so much longer before the Flood than we do today. I began to tackle that question full time with the advent of the new millennium. This little book exists to share what I have found during the past seventeen years of all-out, strenuous research effort on that question.

I have drawn material freely from the pages of *The Biblical Chronologist*, where I published a number of articles on the topic in 2001 and 2002.[5] These I have edited, corrected, and updated as necessary, incorporating them into the early chapters of this book. Later chapters, from Chapter 6 on, contain new material only, published here for the first time.

I have not made much of an effort to popularize my earlier discoveries. This has not been because I think them unimportant or uninteresting. It has been because the research work remaining to be done crowded out everything else.

This must now change. I will be making every effort to popularize the present book, and to commercialize the product it describes so that it becomes readily available globally. Billions of lives depend on it. For this reason, I intend, as a start, to make this book freely available on my website, www.BiblicalChronologist.org. I plan to keep the website

---

[3]Gerald E. Aardsma, *The Exodus Happened 2450 B.C.* (Loda, IL: Aardsma Research and Publishing, 2008). www.BiblicalChronologist.org.

[4]Gerald E. Aardsma, *Noah's Flood Happened 3520 B.C.* (Loda, IL: Aardsma Research and Publishing, 2015). www.BiblicalChronologist.org.

[5]www.BiblicalChronologist.org.

updated with the latest research results, so all can benefit as expeditiously as possible.

But I am hopeful that this book, with its intensely urgent, practical message, will not come to crowd out the research of the earlier books on which it has been built. In point of fact, I think the string of discoveries which began with the seemingly dry and insignificant discovery of a missing thousand years in traditional biblical chronology, are vitally corrective of several pernicious falsehoods which have grown up in the past two centuries in the West. Ideas have consequences, and false ideas generally have bad consequences. We must lay at the feet of these several falsehoods the premature deaths of hundreds of millions of people, as the perceptive reader will glean from this book.

I am afraid that the claims I make in this book will appear impossibly nutty to some readers, despite my best efforts to explain their factual historical and scientific bases. I can only assure you that I am not a nut. My Ph.D. is from a top-flight school, the University of Toronto, and it was earned, with top grades. I deliberately left mainstream academia to pursue my Bible/science interest only a few years after completing my Ph.D. in physics. I am coauthored in a paper with two dozen capable colleagues, including a Nobel Prize-winning physicist, on work I was involved in just prior to my exit.[6] Please take seriously what I am trying to communicate to you in this book. Your life, and the lives of those you love, depend on it.

Gerald E. Aardsma
June 29, 2017
Loda, IL

---

[6] Aardsma, G. et al. Phys. Lett. 194, 321-325 (1987).

Chapter One

Beginnings

> So all the days of Adam were nine hundred and thirty years, and he died. –Genesis 5:5.
>
> So all the days of Methuselah were nine hundred and sixty-nine years, and he died. –Genesis 5:27.
>
> So all the days of Noah were nine hundred and fifty years, and he died. –Genesis 9:29.

According to the record of the ancient past found in the biblical book of Genesis, humans once lived in excess of 900 years. Today, humans rarely live in excess of 100 years. Why were human life spans so very much greater in the distant past than they are today?

I have devoted a substantial portion of my life to this question. My earliest experimental research efforts on it, involving fruit flies and gerbils, were carried out in the basement of a small country house my wife and I rented in the early 1980's. I was in my late twenties at the time, finishing up university as a student of physics. With a growing family to care for, I had little spare time both then and in the next several decades, but I continued to chip away at the human longevity question. By my mid-forties, I had laid sufficient groundwork to mandate that I turn my attention full time to answering it.

I am now in my early sixties. Though research is still in progress and seems likely to continue for some decades, the time has come to share what I now know. It will become increasingly clear through this book that it would be morally wrong of me to do otherwise.

Far more than just curiosity has motivated my efforts to unlock the mystery of human longevity these past four decades. The question of why human life spans were once more than ten times what they are today is one of enormous practical significance. Indeed, there is no medical or scientific question of greater practical significance than this one. It was clear from the start that a correct understanding of why human life spans were so much greater in antiquity might open the door to an understanding of how to extend human life spans today. I was not much concerned about extending my own life span. In our twenties, old age mistakenly seems remote. It was my parents I thought of most often, and ever more urgently as the decades passed.

Dad died in 2000, just a few weeks short of his 77th birthday. He was a good man who had devoted his life to ministry rather than money, with malice in his heart toward no one. The world would be a better place were he still with us.

Mom, now in her late eighties, still retains her natural poise, combined with a New England Yankee reserve and fortitude. I continue to scramble, in hopes that I am not too late to see my youthful dream fulfilled in her.

Cleaning the Slate

Before profitable analysis of the true issues surrounding the ancient mystery of human longevity can begin, there are a number of misconceptions and imprecise definitions in common use which must be cleared up. The meaning of the word "aging" is a good place to begin.

"Aging"

In common use, "aging" can mix together elements of both "maturing" (or "growing up") and "declining" (or "growing old"). Contrary to this common use, biological considerations lead to a natural separation of the concepts of "growing up" and "growing old." "Growing up" is seen biologically as a time of cell proliferation and differentiation. In contrast, "growing old" involves an increasing loss of cell mass and increasing loss of functional ability originating at the cellular level.

Today, humans "grow up" during their first two decades. They then enter a plateau for several decades, during which they are neither maturing physically nor substantially declining. This is followed by another few decades during which physical decline becomes apparent at an ever-accelerating pace, culminating in death.

The phases of a person's life can be likened to the phases in the life of a building. The "growing up" phase corresponds to construction of the building. The plateau phase corresponds to the building's serviceable life. The decline phase corresponds to the building's eventual demise due to loss of structural strength in its materials.

It is natural to separate the concepts of construction and aging when we think about the life of a building. Similarly, the concepts of "growing up" and "growing old" need to be kept separate as we study human aging. The cure for aging will not turn adults back into babies.

In this book, factors affecting the rate of maturation are not of much interest. Factors affecting the length of the plateau and decline phases are the present focus.

The Impact of Modern Medicine on Maximum Life Span

Another common misconception is that people are reaching maximum ages today far in excess of the maximum ages people could hope to obtain a thousand years ago. The popular notion here is that modern science and medicine have brought about a remarkable increase in the maximum length of life.

One has simply to recall the "threescore years and ten" of Psalm 90 (KJV) to know that this idea is false. In point of fact, modern science has been able to accomplish next to nothing to increase the maximum age to which people can live. People have been living into their seventies and eighties and beyond for the past several thousand years. Unfortunately, modern science is totally at a loss at present to know how to extend substantially the maximum human life span—which is why this book is necessary.

What modern science has done is to increase the *average* life span. That is, modern science and medicine have made it possible for a much larger percentage of the population to reach their seventies before dying. For example, in the past, many individuals died in infancy and early childhood as a result of disease. Modern science has found ways to protect children from these diseases, thus enabling many who would have died in infancy in the past to live on into their seventies and beyond in the present. The net effect of this is to increase the *average* age at death for the overall population.

Modern medicine has become very good at keeping people alive long enough for them to reach the decline phase, and modern medicine has become very good at keeping people alive a little longer during the decline phase. It has so far been able to do almost nothing to alter the age at which the decline phase is reached, and it has so far been able to do nothing to alter the inevitability of death within a few short decades of entering the decline phase.

"Special" Groups and Individuals

Still another misconception is that "special" groups or individuals living today have maximum life spans remarkably different from the overall population—either far above or far below the normal life span today. One reflection of this is the notion that "primitive" peoples live only into their thirties.

This is a confusion of average and maximum life spans again. The average life span can be much reduced in primitive living conditions, but this does not alter the maximum possible life span. In primitive living conditions, disease and exposure to a harsh environment can result in the death of many people while they are still relatively young. But one still finds elderly individuals in populations raised in primitive circumstances—individuals who are, in fact, in their eighties and nineties.

Another reflection of this "special groups" idea is the notion that people who live in particular geographical locations (e.g., Tibet) or who hold particular professions (e.g., Tibetan priests) live to extreme ages. In actual fact, no dependence of maximum life span on geographical location or profession is found when authenticated records of individuals of verifiable identity are examined.

"Normal" Life Span

Perhaps the most difficult misconception to correct is the belief that 75 years is a normal life span for humans. The Genesis record of human life spans—Noah living to 950, for example—shows immediately that this belief is simply false. If we are to take biblical history seriously, then we must conclude that death at 75 is not normal for humans at all.

Imagine an island community, cut off from the rest of the world, where everybody dies before age 40 due to a certain doubly recessive genetic defect which has come to be found in all individuals in the population.

This defect causes them all to be highly susceptible to cancer. As a result, all contract cancer and die in their fourth decade of life.

If this community remained cut off from the rest of the world for many generations, it is easy to see how they could ultimately come to believe that death by age 40 was normal for humans—and not only normal, but indeed proper. It is probable that many of them would respond with skepticism and disdain if someone were to suggest the idea that many of their distant ancestors, who had discovered and populated the island thousands of years previously, had lived into their eighties and nineties. Certainly many of them would find the suggestion incredible that practical steps (i.e., marriage outside the island population) might be taken to restore an average life span of 75 years to their community. And some, no doubt, would assert that it was the will of God for humans to die before age 40, and that it was impious to meddle in such matters.

But they would be wrong.

The Genesis longevity data teach us that the world in which we live today is like this island community. Seventy-five years has become the average life span. It has been this way for thousands of years. But it is entirely wrong to mistake that to which we have become accustomed for that which should rightfully be.

A New Hypothesis

Genesis teaches us that we must reorient our thinking. We must recognize that the present human life span of 75 years is a very sad state of affairs indeed. Much more dramatic than our imaginary islanders whose life spans were reduced a mere factor of two, our life spans have been reduced by over a factor of ten. Far from 75 years being "normal" for humans, we must acknowledge that the entire human population today is, in fact, subject to a devastating malady.

This idea, that human aging, as we know it, is a malady—a disease— is the fundamental hypothesis underlying this book. You will find that this hypothesis has been corroborated beyond reasonable doubt by the time you have finished reading this book.

We have learned to call this malady "old age," and we have learned to accept it. But Genesis shows us that this is entirely wrong-headed. It shows us that "old age" is a false label, and a highly misleading one. When we come to grips with what Genesis plainly shows and accept it at face value, we see immediately that nobody has ever died of "old age" at 75 or even at 125. The Genesis life span data teach us that 75 is not an *old* age. It is laughable to call an individual "old" who has lived only 75 years in a population sporting many individuals in excess of 750 years, as was the case in the long-ago days recorded in the early chapters of Genesis.

The biblical life span data make it clear that nobody dies of "old age" at 75 years, for 75 years is *not* an *old* age for humans at all. People routinely die within a few decades of the *young* age of 75 today, but they do not die *because* of their age. Time is not the killer. They die because they have been afflicted with a devastating malady which tends to kill humans within a few decades of 75 years today. This malady decimates our bodies, causing them to lose functional ability and waste away *while we are still very young*—before we have achieved even one tenth of our life span potential.

Malady $\bar{X}$

To focus attention on the true essence of the longevity problem, I will, from this point on, use "aging" and "old age" only in quotes, and I will make use of the new term, "malady $\bar{X}$" (read "malady X-bar"). For example: "He died of malady $\bar{X}$" rather than "He died of old age." (The logic behind this nomenclature will be explained in a subsequent chapter.)

By substituting "malady $\bar{X}$" for "old age," I mean deliberately to part company with the false idea that people die within a few decades of 75 today because they are *aged* and replace it with the true idea that people die within a few decades of 75 today because they are *afflicted*. By putting "aging" and "old age" in quotes, I mean to make it perfectly clear that *time* is not the essence of the problem. I mean to emphasize that the essence of the problem is what medicine routinely calls *disease*.

If we are to think accurately about longevity in light of what Genesis shows us, then we must begin to see what we presently call "old age" as simply another human disease. I have introduced the new term "malady $\bar{X}$" as a temporary name for this disease to facilitate its study until we have learned enough about it to name it something more appropriate.

Malady $\bar{X}$ is a disease that manifests itself by, among other things: loss of hair color, wrinkled skin, vision impairment, loss of physical strength, and increasing susceptibility to a large number of diseases. Malady X̄ symptoms are universally seen in all individuals over the entire globe today beginning in the fifth decade of life. The sad result is death of most individuals within a few decades of 75 years of age, and of all individuals before 130 years of age—dramatically short of the known life span potential of humans, in excess of 900 years.

From the beginning, my research problem was to find the physical cause of malady X̄. The hope and expectation of this research was that once the cause of malady X̄ had been found, a cure would be able to be formulated. The further hope and expectation was that, once a cure for malady X̄ had been formulated and appropriated, the symptoms of malady X̄ would no longer appear in any individual's fifth decade, and people would be able to go on living in the plateau phase of life for hundreds of years, just as they did back in Genesis.

The Number One Health Problem

Having clarified the fundamental essence of the longevity problem, it is possible to correct another common misconception. This is the idea that killer diseases such as cancer and cardiovascular disease are mankind's primary health problems today. In actual fact, malady X̄ is the primary health problem.

Cancer and cardiovascular disease are, for the most part, diseases of "old age." That is, they prey on individuals already weakened by malady X̄. The implication is that the incidence of cancer, cardiovascular disease, and all other "old age" related diseases will dramatically decline once the cure of malady X̄ has been appropriated.

Note that the converse is not true. Even if total cures for cancer and cardiovascular disease were to be found, people would still continue to die of malady X̄ within a few decades of 75 years, due to diabetes, or Alzheimer's, or pneumonia, or...

A cure for malady X̄ is clearly, by far, the most pressing medical need today. All other diseases *combined* pale in significance relative to the misery and suffering caused each year by malady X̄.

The Difficulty of the Research Problem

The magnitude of the problem of malady X̄ and the urgency of its solution has long been recognized. But finding a cure for malady X̄ has proven to be no easy task. Despite billions of dollars spent on research, modern science is presently at a complete loss regarding how human life spans might ever be significantly increased beyond 100 years. Some well-respected scientists even claim that it is a fundamental impossibility.

> Leonard Hayflick, an expert on aging at the University of California, San Francisco, denounced what he called "outrageous claims" by some scientists that humans are capable [of] living well past 100 years.
>
> "Superlongevity," he said, "is simply not possible."[7]

The apparent intractability of the problem is underscored by consideration of present life span statistics. Despite a current global population of seven and a half billion people—implying in excess of one hundred fifty thousand deaths per day worldwide (i.e., fifty five million deaths per year)—a life span in excess of 120 years is still a rare and remarkable event, and not one verifiable case of any individual living past 130 years of age has ever been found in modern times.

The difficulty of the problem is further emphasized by the fact that, while the scourge of reduced longevity has been with us for over five thousand years, no one in all that time has been able to discover how to do anything about it.

The biggest difficulty for the modern researcher has been that *everybody* suffers from this disease today. Normally a researcher studies a group of diseased individuals relative to a group of healthy individuals. In the case of malady $\bar{X}$, there are no healthy individuals to compare to.

If even one individual were to have lived in modern times to, say, 150 years, that individual would surely have been the subject of intense scientific interest. The interest would have focused around the question of what factor or factors had allowed that individual to live so long. Every effort would have been made to isolate factors in that individual's experience which had differed from everybody else, with the expectation that one or more such factors would be found to be responsible for the difference in longevity observed.

But we have no such individual or group of individuals to compare to today. *Everybody* is afflicted with malady $\bar{X}$. The search for differences has no subjects from which even to begin.

---

[7]San Francisco (AP), "Life expectancy may be nearing its upper limit," *The News-Gazette* (Champaign-Urbana, Illinois), 19 February 2001, p. A-1 and A-6.

A Lone Hope

Today, that is.

The search for a cure for malady $\bar{X}$ does have a few subjects to work with from the distant past, if we are willing to believe Genesis: Adam, and Noah, and Arpachshad, and Peleg, and Abraham, for example. Genesis tells us plainly that these men all enjoyed life spans well in excess of 150 years. Is it inappropriate or silly to try to isolate one or more factors in their experience which may have differed from our experience today? It seems to be our sole hope.

Conclusion

All investigators admit that the problem of how to extend human life spans is one of extreme difficulty. Reliable data from subjects living beyond even 150 years—the sort of data one really needs to have any serious hope of cracking the problem—cannot be obtained today. Many researchers have already spent much time groping about in the dark for some clue to the mystery of human longevity. Unfortunately, they have nothing to show for their efforts. Millions of individuals continue to die each year, most before they have lived even 80 years, as has been the case for thousands of years.

Only one soft ray of light transgresses this blackness. It glimmers unobtrusively but faithfully from a lone window which looks out dimly upon an ancient world where thousands of multi-centenarians once worked and played. I suggest the time may have come to take a careful look through this window. It seems to be our only possible hope. And perhaps it was put there for this very purpose.

CHAPTER TWO

THE DATA

The present work breaks with other contemporary scientific research on life spans in its attitude toward the Genesis life span data. These data are commonly held to be mythological by contemporary researchers. The attitude toward these data underlying the present work is opposite to this. I hold these data to be valid, accurate, historical observations of actual life spans of real individuals.

This attitude is neither arbitrary nor religiously biased. The idea that these data are mythological or otherwise concocted cannot be retained by any scientist who has actually worked with them. These biblical data display certain features which are impossible to explain in any other way than that they are valid historical observations. This property will become increasingly clear as we proceed through this book. For now, I simply point out that the basically historical nature of these life span data is already strongly implied by their intimate association with key biblical chronological data. Many of these life span numbers are recorded together with birthdate numbers which are used directly in the construction of the chronology of biblical history stretching back to Adam. When this biblical chronology is checked using radiocarbon dating, it is found to be remarkably accurate. For example, the biblical chronology date for Noah's Flood is 3520±21 B.C., and the corresponding radiocarbon date is 3525±12.5 B.C.[8] This is probably the most secure and precise date humankind possesses for any historical event of such remote antiquity. The biblical chronological numbers are demonstrably historical. Would it not be odd if the closely associated life span numbers were mythological?

Once the biblical life span data are accepted as historical and reliable, they automatically become the focus of our research interest. They

---

[8]Gerald E. Aardsma, *Noah's Flood Happened 3520 B.C.* (Loda, IL: Aardsma Research and Publishing, 2015), 307–313. www.BiblicalChronologist.org.

do so because they report on a unique, real-life, natural "experiment" which displays a pronounced life span effect in humans. This is the *only* experimental evidence we have that maximal human life spans can be altered. It is essentially certain that this "experiment" will never be repeated, both because it covers many generations over thousands of years, and because deliberate scientific experimentation of this sort on humans would be blatantly unethical. Thus, not only are the biblical data the sole experimental data we *presently* have which show anything of interest regarding human longevity, but also they are almost certainly the sole experimental data of the sort on humans we will *ever* have.

Furthermore, experimental data displaying *any* evidence for extension of the present human life span are obviously of extreme interest. But the Genesis longevity data go far beyond this, giving clear evidence of more than a *tenfold* increase in life spans.

Simply stated, the biblical life span data record a very pronounced life span alteration "signal" in humans. No other data anywhere record any life span alteration "signal" in humans at all. Obviously, the biblical data make easy claim to our entire attention.

Life Spans

The biblical life span data of interest to the present study are shown in Table 1. The ages at death are taken from the Bible, from the verses shown in the "Bible Reference" column.

Though this list is comprehensive, it is not exhaustive. To begin with, biblical individuals with anomalously low life spans, such as Enoch (Genesis 5:24), Lamech (Genesis 5:31), and Nahor (Genesis 11:24–25) have been excluded. Secondly, biblical females have also been excluded. Males and females have different average life expectancies, so they need to be treated separately. There are significantly more life span data points for biblical males than there are for biblical females, so the present study focuses exclusively on male life spans, both ancient and modern. Finally, no attempt has been made to add data to the list after Moses. Such data are of limited interest in the present study. They show mainly a continuation of the 75-year average life span which, on the basis of Psalm 90 ("A Prayer of Moses the man of God") was already operative by the time Moses died.

Table 1: Selected biblical life span data.

Name	Age at Death	Bible Reference
Adam	930	Genesis 5:5
Seth	912	Genesis 5:8
Enosh	905	Genesis 5:11
Kenan	910	Genesis 5:14
Mahalalel	895	Genesis 5:17
Jared	962	Genesis 5:20
Methuselah	969	Genesis 5:27
Noah	950	Genesis 9:29
Shem	600	Genesis 11:10–11
Arpachshad	438	Genesis 11:12–13
Shelah	433	Genesis 11:14–15
Eber	464	Genesis 11:16–17
Peleg	239	Genesis 11:18–19
Reu	239	Genesis 11:20–21
Serug	230	Genesis 11:22–23
Terah	205	Genesis 11:32
Abraham	175	Genesis 25:7
Ishmael	137	Genesis 25:17
Isaac	180	Genesis 35:28
Jacob	147	Genesis 47:28
Levi	137	Exodus 6:16
Joseph	110	Genesis 50:22, 26
Kohath	133	Exodus 6:18
Amram	137	Exodus 6:20
Aaron	123	Numbers 33:39
Moses	120	Deuteronomy 34:7

Despite these exclusions, the list presents a remarkable data set. Its 26 data items clearly capture the past decline in human longevity.

Birthdates

In the present study, the Table 1 life span data are everything. Understanding why they declined as they did is the goal. It is not sufficient simply to observe *that* life spans were longer in the past. The goal is to discover *why* they were longer—what physical, material agent(s) caused human life spans to shorten. The goal is to elucidate a cause and effect relationship. To accomplish this, knowledge of how life spans changed *with time* is needed. Thus, to make full use of these life span data, it is necessary to attach a unique time to each data item, specifying when in history that particular life span applied. This can be accomplished by assigning proper calendrical birthdates to each of the individuals listed in the table. If it is then assumed that each of the individuals shown in the table died of malady $\bar{X}$, then the life spans of these individuals can be used as an estimate of the malady $\bar{X}$-specific life expectancy when they were born. This will allow a graph to be constructed of life expectancy versus time for the ancient past, which is what is needed to attack the problem of the cause of reduced human longevity quantitatively.

Table 2 shows the needed birthdates. These dates have been computed from a combination of both biblical and extra-biblical chronological data according to the principles of the modern discipline of biblical chronology. The "Chronology Numbers" column shows numbers which are needed in the computations of the birthdates, and the "Bible Reference" column shows where these numbers were obtained.

The table includes two chronological reference points: Noah's Flood at 3520 B.C.,[9] and the Israelite Exodus from Egypt at 2447 B.C.[10] The date of the Flood is used as the reference point for most of the dates in the table. But chronological continuity, provided by recording the age of the father at the birth of the son, is lost following Jacob, during the 450 years Israel was in Egypt, making the final six birthdates of the table somewhat more complicated to compute.

---

[9]Gerald E. Aardsma, *Noah's Flood Happened 3520 B.C.* (Loda, IL: Aardsma Research and Publishing, 2015). www.BiblicalChronologist.org.

[10]Gerald E. Aardsma, *The Exodus Happened 2450 B.C.* (Loda, IL: Aardsma Research and Publishing, 2008). www.BiblicalChronologist.org.

Table 2: Birthdates of selected biblical males.

Name	Date (B.C.)	Chronology Numbers	Bible Reference
Adam	5176	130	Genesis 5:3
Seth	5046	105	Genesis 5:6
Enosh	4941	90	Genesis 5:9
Kenan	4851	70	Genesis 5:12
Mahalalel	4781	65	Genesis 5:15
Jared	4716	162+65	Genesis 5:18, 21
Methuselah	4489	187+182	Genesis 5:25, 28
Noah	4120	600	Genesis 7:11
Shem	3617	100	Genesis 11:10
—Flood—	3520	1	lasted 1 year
Arpachshad	3517	2	Genesis 11:10
Shelah	3482	35	Genesis 11:12
Eber	3452	30	Genesis 11:14
Peleg	3418	34	Genesis 11:16
Reu	3388	30	Genesis 11:18
Serug	3356	32	Genesis 11:20
Terah	3297	30+29	Genesis 11:22, 24
Abraham	3167	205-75	Genesis 11:32; 12:4 and Acts 7:4
Ishmael	3081	86	Genesis 16:16
Isaac	3067	100	Genesis 21:5
Jacob	3007	60	Genesis 25:26
Levi	2943	27	estimated
Joseph	2916	91	Genesis 41:46; 45:6; 47:9
Kohath	2806	138	estimated
Amram	2668	138	estimated
Aaron	2530	83	Exodus 7:7
Moses	2527	80	Exodus 7:7
—Exodus—	2447		

The birthdate of Joseph can be calculated using the three Bible references provided for Joseph in the table. These reveal that Joseph was 39 years old when Jacob was 130 years old. Thus Jacob was 91 years old at the birth of Joseph.

I have estimated Levi's date of birth from Joseph's date of birth by making use of the fact that Joseph was the ninth son after Levi. Natural child spacing tends to be about three years (compare Moses and Aaron, for example) yielding 27 years between Levi and Joseph. This simple means of estimation is called into question in this instance by the fact that four separate wives were involved. I have chosen to ignore this potential complication since, from the primary record found in Genesis 29 and 30, the births seem to have been consecutive rather than overlapping, and because even a crude estimate will suffice for the present purpose.

For calculating the dates of birth of Moses and his brother Aaron, the date of the Exodus must be used. Exodus 7:7 reveals that Moses and Aaron were 80 and 83 years old respectively when they confronted Pharaoh. This places the birth of Moses in 2527 B.C.

The birthdates of Kohath and Amram can only be estimated. Even a crude estimate will suffice here once again, because life spans were not changing very rapidly when Kohath and Amram lived. I have simply placed Kohath and Amram equally apart in the 413-year interval spanned by the father-son lineage, Levi–Kohath–Amram–Aaron.

Life Expectancy

Life expectancy is the length of time a person may be expected to live. Since the current study is concerned with human longevity, it is the life expectancy at birth which is of interest. For the current study, life expectancy at birth data are obtained by combining the age at death data from Table 1 with the birthdate data from Table 2. The result is shown in Table 3.

Normally, life expectancies are calculated by averaging ages at death from a large population. The biblical life span data provide only individual data points, not averages over many individuals. This is the same as taking the age at death of a single modern male to estimate the life expectancy at birth of modern males worldwide. While the average life expectancy is likely to be near 75 years at present, not all males die at age 75 years. Instead, ages at death of, for example, 62 years or 87 years are

Table 3: Point estimates of life expectancies from Adam to Moses.

Name	Date (B.C.)	Life Expectancy
Adam	5176	930
Seth	5046	912
Enosh	4941	905
Kenan	4851	910
Mahalalel	4781	895
Jared	4716	962
Methuselah	4489	969
Noah	4120	950
Shem	3617	600
Arpachshad	3517	438
Shelah	3482	433
Eber	3452	464
Peleg	3418	239
Reu	3388	239
Serug	3356	230
Terah	3297	205
Abraham	3167	175
Ishmael	3081	137
Isaac	3067	180
Jacob	3007	147
Levi	2943	137
Joseph	2916	110
Kohath	2806	133
Amram	2668	137
Aaron	2530	123
Moses	2527	120

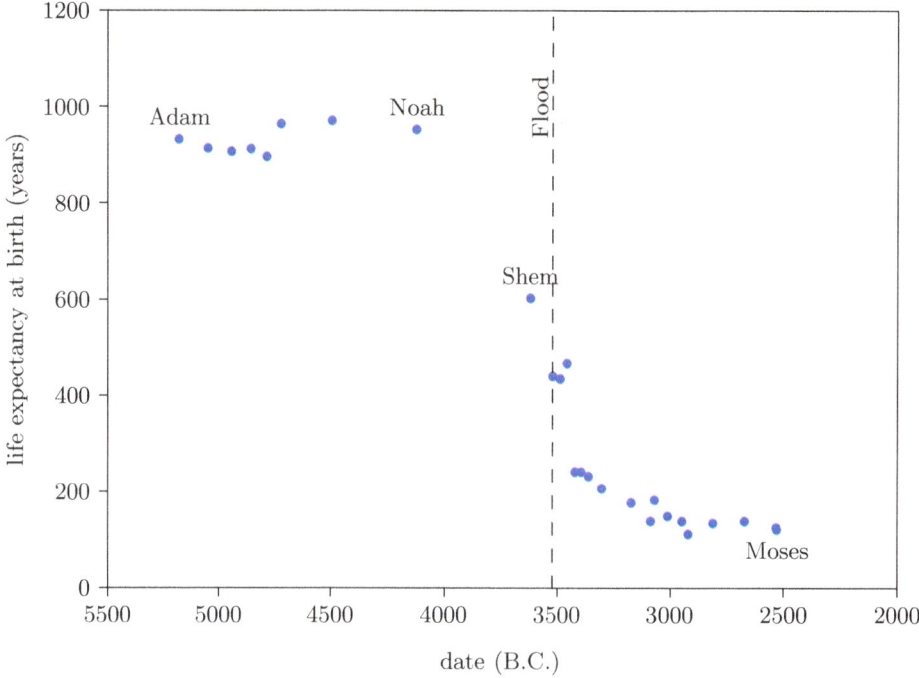

Figure 1: Biblical life expectancy at birth data for selected males.

quite common. Individual ages at death today can easily differ from the average by plus or minus twenty years. The same is true of the biblical life expectancy data points. In fact, the average life expectancy at birth of the first seven individuals in the table, taken from a time interval during which life expectancies are thought to have been stable, is 926 years, with a standard deviation of plus or minus 28.9 years.

Conclusion

Figure 1 shows a graph of these biblical data. Take a good look at it. There is no graph in the whole world of greater practical humanitarian importance than this one. It is the key which unlocks the mystery of the cause, and of the cure, of "aging," as subsequent chapters will show.

CHAPTER THREE

THE BIBLICAL LIFE EXPECTANCY GRAPH

Before beginning to apply the data displayed in Figure 1, it seems appropriate to point out two unique advantages from which the present research has benefited.

The first unique advantage has been the ability even to construct this graph.

To construct this graph, one has to have one's biblical chronology right all the way back to Adam. Most importantly, as will become increasingly clear, one has to have the date of the Flood right. The key to these prerequisites is the recognition that traditional biblical chronology has dropped out a full millennium in 1 Kings 6:1.[11] This discovery was made only in 1990. Extension back to Adam of the new biblical chronology resulting from that discovery was not completed until 1999.[12] Thus, though the biblical life span data of concern to this study are of great antiquity, ability to plot them accurately on the graph of Figure 1 is less than two decades old.

In addition to having biblical chronology right, the research reported in this book enjoyed a second important advantage. Figure 1 shows plainly that Noah's Flood is the dividing line between the short life span regime of the present day and the long life span regime of the ancient past. The reduced life expectancy data point corresponding to Shem, just before the Flood, may seem to be an exception to this, but Shem's reduced life span results from the fact that he lived most of his life in the post-Flood period. He was born 97 years prior to the Flood, and he died 503 years after the Flood.

---

[11]Gerald E. Aardsma, *A New Approach to the Chronology of Biblical History from Abraham to Samuel*, 2nd ed. (Loda, IL: Aardsma Research and Publishing, 1993).

[12]Gerald E. Aardsma, "A Unification of Pre-Flood Chronology," *The Biblical Chronologist* 5.2 (March/April 1999): 1–18.

The observation that Noah's Flood is the dividing line between the ancient and modern longevity regimes implicates Noah's Flood as the fundamental *cause* of reduced life spans today. Noah's Flood appears to have done something which subsequently shortened human life spans.

What exactly did it do? This is the fundamental question which must be answered in seeking to solve the cause of malady $\bar{\text{X}}$.

Clearly, to have any hope of answering this question, an accurate idea of the nature of the Flood is needed. Indeed, the paramount importance of an accurate knowledge of the nature of the Flood to cracking the longevity mystery will become increasingly apparent in subsequent chapters. For now, the point to notice is that the true nature of the Flood was discovered only in 1997.[13]

Thus, both of the ingredients needed to make quantitative, scientific sense of the biblical life span data—a correct biblical chronology and a correct understanding of the nature of the Flood—have become available only in the past two decades.

These two unique advantages did not guarantee, of course, that this ancient mystery could, at long last, be solved. But they did make it perfectly clear that the time had come for a new, all-out assault on the problem.

A Powerful Instrument

The Figure 1 graph (page 34) is a powerful research instrument. Consider its application in the following three cases, each dealing with common theories of "aging."

Supernatural

Perhaps the simplest lay theory of "aging" is that the human life span is determined by God in ways that cannot be understood or ascertained by humankind. This denies any natural cause of "aging," which immediately yields the corollary that scientific investigation into the matter is useless.

The biblical life expectancy data argue strongly against this theory. They show that human life spans declined in a fairly smooth way from roughly 925 years on average before the Flood to roughly 75 years on

---

[13]Gerald E. Aardsma, "The Cause of Noah's Flood," *The Biblical Chronologist* 3.5 (September/October 1997): 1–14.

average following the Flood. This smooth decline took more than a thousand years to settle out to the currently active average life span. If the human life span is fixed by God, then these data require that God performed numerous miracles, continuously readjusting human life spans for more than a thousand years following the Flood. This seems severely contrary to what we learn of the nature of God's supernatural activity elsewhere in the Bible. Based on the miracles we read about in the Bible, such as the conversion of water into wine (John 2:1–11), or the calming of the sea (Luke 8:22–25), or the iron axe head which was made to float on water (2 Kings 6:1–7), we expect miracles to be generally evidenced as point-in-time suspensions of the natural order, not as innumerable slight adjustments of the natural order.

Meanwhile, we expect natural processes to change smoothly with time. For example, the temperature of a bowl of hot water naturally changes in a smooth progression from hot to cold with time (Figure 2).

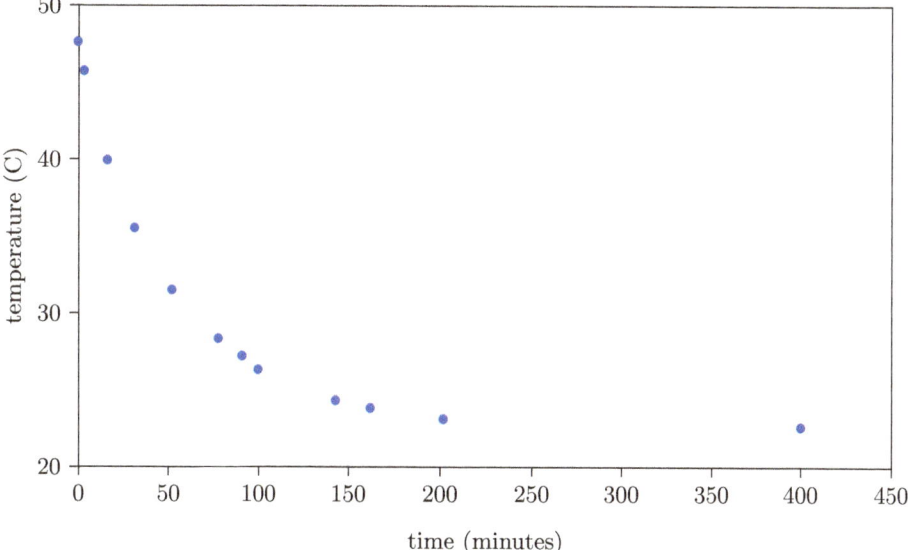

Figure 2: Measured temperature versus time for a bowl of hot water.

Furthermore, the change in temperature of the hot water is more rapid at first, and slows as it nears room temperature. This property is seen in the biblical life span data as well; the rate of change of life spans is rapid immediately following the Flood, and slows as the present value near 75 years is approached.

The biblical life span data reveal that life spans declined in a natural way following the Flood, implying that some natural cause was responsible for this decline.

Vapor Canopy

A second lay theory, held by some who regard Genesis as historical but who misunderstand the history Genesis records, is that pre-Flood longevity was due to a water vapor canopy which enveloped the earth prior to the Flood. This canopy was supposedly suspended above the atmosphere before the Flood, but it condensed and fell to the earth as rain at the time of the Flood, thereby contributing to the Flood's forty days and nights of rain.[14]

The canopy is credited by its adherents with prolonging life prior to the Flood, usually in one of two ways. The first is through attenuation of hypothetically harmful radiation from space. Some versions of the theory cite ultraviolet rays from the sun, others cite cosmic radiation. The second is through enhanced atmospheric pressure due to the weight of the vapor canopy on the atmosphere.

Both of these versions of the canopy/longevity theory are immediately falsified by the biblical life expectancy data. To see this, notice that any attenuation of harmful radiation would immediately have ceased upon collapse of the canopy at the time of the Flood. Similarly, atmospheric pressure would have changed suddenly and completely upon condensation of the canopy at the time of the Flood. Thus, human life spans should have changed to their post-Flood value suddenly and completely at the time of the Flood. But the biblical data show us that life spans did not change suddenly and completely at the time of the Flood. Rather, they took more than 1000 years to complete their change from

---

[14]I have never seen an explanation of how such a canopy of water vapor would be kept in place and doubt that one can be found. Most difficult to understand is what would keep the water molecules from mixing with the rest of the atmosphere. I know of no way to accomplish such a thing. Notice that the atmosphere today does a very good job of mixing all of its constituents together. We do not find separate layers of oxygen, nitrogen, carbon dioxide, water vapor, or any other gas. Other serious scientific problems a vapor canopy introduces include greenhouse heating of the surface of the earth and inordinate heating of the atmosphere at the time of the Flood due to the heat of condensation of water vapor and the conversion of gravitational potential energy to heat energy which would result from collapse of such a canopy. There seems to be little scientific justification for continued adherence to the idea of such a canopy.

the pre-Flood value of 925 years to the present value of roughly 75 years. Canopy/longevity theories may be safely discarded.

Evolution

A broad spectrum of scientific theories about "aging" falls under the general umbrella of "evolutionary." The central idea in these theories is that "aging" is a by-product of evolution.

One of these theories suggests, for example, that all evolution needs is propagation of the species, and once this function has been fulfilled, an organism is best gotten rid of so it doesn't use up valuable resources. Thus, evolution has arranged for organisms to be discarded once their reproductive task has been completed.

Notice that the entire category of such theories is falsified by the biblical life expectancy data. The idea that a species' longevity is somehow determined by its evolutionary history—specifically, that humans live to 75 years on average because they have been somehow programmed by evolution to do so—cannot be true, because the biblical life expectancy graph shows that humans lived in excess of 900 years only a few thousand years ago. The biblical life expectancy data, in fact, falsify all theories of human longevity which hold death within a few decades of 75 years to be a pre-programmed biological necessity.

Evidently, what we call "aging" today really has nothing to do with evolution at all. The biblical life expectancy graph implies instead the radically new idea that what we call "aging" today has everything to do with catastrophe-occasioned disease.

Conclusion

Very many theories have been advanced in an effort to explain why humans "age" and die the way we do. Theories multiply in science whenever none of them really works. The biblical life expectancy graph easily falsifies many existing theories, clearing the fog, while pointing in a totally new direction.

Chapter Four

The Central Hypothesis

Years of contemplating the nature of "aging" in light of the biblical life expectancy data led me ultimately to conjecture that malady $\bar{X}$ must be a nutritional deficiency disease. This conjecture rapidly became my central hypothesis.

Deficiency Diseases

The human body is made up of billions of microscopic cells. Each cell can be thought of as a very complex and busy city, part of a vast empire (the body). Each moment, raw materials flow into these busy cities, and, together with some waste, many finished products necessary to the overall growth, function, and maintenance of the empire flow out.

Among the raw materials flowing into these cities each moment are some which can be obtained only from outside the boundaries of the empire. (For example, the human body cannot manufacture oxygen. The human body gets oxygen by breathing it in from the atmosphere.) Many of these raw materials are absolutely vital to the cities—the cities cannot produce necessary finished products without them. If the supply of any one of these vital raw materials is halted for any reason (for example, lack of oxygen due to asphyxiation), production of one or more vital finished products ceases. The health of the empire then suffers and, if the lack of this vital raw material persists long enough, the empire eventually disintegrates (i.e., the body dies).

On the list of vital raw materials needed by our bodies are such things as oxygen, water, carbohydrates, fats, amino acids, certain minerals—calcium, phosphorus, sodium, potassium, chloride, magnesium, iron, copper, iodine, and many others in minute amounts—and a curious assortment of just over a dozen organic substances we call vitamins. If for any reason the cells of the body are unable to obtain one of these vital substances, then a deficiency disease results.

The most common cause of deficiency diseases in humans is inadequate diet. The essential raw material is simply not being taken into the body. But there are other possible causes, such as a faulty digestive system resulting in inadequate absorption of an essential raw material once it has been ingested, or combination of the essential raw material after ingestion with some other chemical and subsequent elimination of the compound from the body.

Of the list of essential raw materials needed by the body, the vitamins are of particular interest in the present context. Malady $\bar{X}$ appears to be a vitamin deficiency disease.

The Example of Vitamin C

Scurvy is an example of a vitamin deficiency disease. It results from a diet deficient in vitamin C.

Before it was understood that scurvy is a deficiency disease, scurvy was a common disease of mariners. Vitamin C is abundant in fresh fruits and vegetables, so most of us get plenty of it in our normal diet each day. Vitamin C is easily subject to destruction by oxidation, however, so vitamin C levels decline in fruits and vegetables following harvest. Upon prolonged storage, vitamin C levels in fruits and vegetables become inadequate to meet human dietary requirements for this substance. The difficulty of providing mariners with fresh produce on long sea voyages inevitably resulted in many cases of scurvy.

Long before vitamin C was discovered, a number of individuals began to understand that scurvy could be prevented by a diet containing adequate fresh fruits and vegetables. Ways were sought, and eventually found, to protect the anti-scurvy property of lemons by concentrating and preserving the juice. Early in the 1800's, the British navy adopted regulations requiring daily consumption of lemon juice, bringing the scurvy plague to an end in the British navy. Eventually, this simple remedy was adopted by commercial vessels as well. The substitution of cheaper lime juice for the original lemon juice led eventually to the slang designation of British sailors as "limeys."

The actual anti-scurvy factor in fresh fruits and vegetables—the vitamin C molecule—was isolated, and its structure determined (Figure 3),

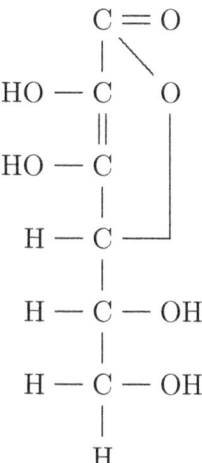

Figure 3: Structure of the vitamin C molecule.

only about eighty years ago.

Vitamin C is a relatively simple organic molecule, but the human body is unable to synthesize it. This simple molecule is vital to human health. Without it, connective tissues between cells degenerate. This results in a complex of symptoms at the whole-organism level. Most conspicuously, blood vessels become so weak that hemorrhage results, and teeth lose their strength and become diseased.

> Adult patients suffering from scurvy complain of weakness, pains in their legs, swollen and bloody gums and hemorrhages. Examination discloses petechiae, chiefly about the hair follicles of the lower extremities and sometimes brawny, tender thighs. All of these features are due to hemorrhage...
>
> Weakness is usually the first thing complained of by persons suffering from vitamin C depletion. Fatigue, palpitation and breathlessness are also common. The patients dislike to stand or walk and often affect a rather characteristic standing position with their legs slightly flexed. The complexion is pallid and dirty looking. Gingivitis occurs, followed by loosening of the teeth, a consequence of resorption of the alveolar bones and infections about the teeth and is accompanied by a foul breath. Other signs of scurvy are hematuria, bloody diarrhea, nasal hemorrhage or hematomas about the jaw or

bones of the lower extremities.[15]

Vitamin C is needed only in minute amounts—about one ten-thousandth of our daily food intake on a dry weight-per-weight basis. This miniscule daily requirement relative to the bulk diet is characteristic of all the vitamins. In the case of vitamin D, the amount needed is roughly one five-millionth of our daily food intake. But though so little is needed, this small amount is absolutely essential. Without it, our cells lose their ability to carry out their jobs, and, eventually, a complex of whole-body symptoms—a deficiency disease—develops.

Central Hypothesis

My central hypothesis is that "aging" is a vitamin deficiency disease resulting from dietary insufficiency of a previously unknown vitamin which was made globally deficient by the impact of Noah's Flood on earth's environment.

For many years, before I knew what this vitamin was, I called it simply "vitamin X." It was then logical to call the disease one gets when one is deficient in vitamin X, "malady $\bar{X}$" (malady X-bar). The "bar" signifies negation. "Malady $\bar{X}$" means "the disease due to not X" or "the disease due to lack of X." In this nomenclature, scurvy is "malady $\bar{C}$," beriberi is "malady $\bar{B}_1$," rickets is "malady $\bar{D}$," and "aging" is "malady $\bar{X}$."

Early support for this hypothesis was easily gleaned by comparing and contrasting malady $\bar{X}$ with scurvy.

Complex of Symptoms

Notice, first of all, that like scurvy, malady $\bar{X}$ ("old age") exhibits itself as a complex of whole-body symptoms: skin loses its elasticity, muscles weaken and decrease in size, hair loses its color and thins out, bones become brittle, eye lenses stiffen...

These are very diverse symptoms, yet they show up together in "old age." One could suppose that they are all caused by independent physiological malfunctions of one sort or another, and that these independent

---

[15] Walter H. Eddy and Gilbert Dalldorf, *The Avitaminoses: The Chemical, Clinical and Pathological Aspects of Vitamin Deficiency Diseases* (Baltimore: The Williams & Wilkins Company, 1937), 175.

malfunctions are all synchronized by some sort of master biological timeclock. But much simpler is the idea that these diverse symptoms are simply varied macroscopic manifestations of a single missing essential molecular component at the microscopic, cellular level—just as is the case with scurvy.

Particular Symptoms

Not only is there a complex of whole-body symptoms in both cases, but some of the particular symptoms of "old age" also show striking similarities to symptoms of scurvy.

> Aschoff and Koch were greatly impressed with the similarity of the scorbutic [scurvy] lesions to those in senility. The changes in cortical bone are difficult, if not impossible to distinguish. ...In both conditions the bones are notably thin and rarefied, susceptible to fracture and defective in the ability to form a callus once fracture has occurred. ...
>
> Westin interpreted the tooth lesions as similar to the atrophy of old age and said scurvy may be considered to hasten involution. In his cases the teeth showed the same resistance to caries that is seen in senility as well as the rarefaction common to advanced years.[16]

This demonstrates that vitamin C deficiency disease can produce precisely the same sorts of abnormal changes and injury to body tissues as those which are characteristic of "old age." Evidently then, at least some of the specific symptoms accompanying "old age" fall naturally within the vitamin deficiency disease category.

Apparent Contrast

An apparent contrast between "old age" and scurvy is that only a small percentage of individuals in a normal population ever contract scurvy, while all individuals, if they live long enough, contract malady $\bar{\mathrm{X}}$.

---

[16]Walter H. Eddy and Gilbert Dalldorf, *The Avitaminoses: The Chemical, Clinical and Pathological Aspects of Vitamin Deficiency Diseases* (Baltimore: The Williams & Wilkins Company, 1937), 194.

This apparent difference is easily explained. Normal diets of most individuals supply them amply with vitamin C. Only the few individuals on deficient diets ever contract scurvy. In contrast, normal diets of *all* individuals have become seriously deficient in vitamin X as a result of the Flood. How this came about will be explained in detail in subsequent chapters.

Variable Time of Onset

Another similarity between "old age" and scurvy is that the time of onset can be varied. Prior to the Flood, men lived in excess of 900 years before they succumbed to "old age" and died. After the Flood, men contracted "old age" at younger and younger ages, until the present, much-diminished life span near 75 years was reached.

The time of onset of scurvy can be similarly varied:[17]

> They found that less than 50 cc. of milk daily resulted in scurvy within thirty days, that 50 cc. delayed the onset of the disease until the seventy-fifth day and that 100–150 cc. of milk postponed evidence of scurvy for four months.

Why Life Spans Changed

Milk is a poor source of vitamin C. Thus all of the animals (guinea pigs) referred to in the previous quote were subject to a vitamin C deficient diet. Those getting less milk got less vitamin C. Thus the time of onset of scurvy is seen to be directly related to the daily dose level of vitamin C in the diet.

This dose-dependent time-of-onset characteristic of vitamin deficiency diseases explains the change in human life spans following the Flood which Genesis records. Earth's environment uniformly meters out vitamin X to all individuals globally in a rigidly fixed dose from the time of birth on. Life spans diminished following the Flood because vitamin X, which was already somewhat deficient before the Flood (resulting in death due to "old age" at roughly 925 years), became increasingly scarce globally after the Flood. By 2500 B.C. (one thousand years after the

---

[17] Walter H. Eddy and Gilbert Dalldorf, *The Avitaminoses: The Chemical, Clinical and Pathological Aspects of Vitamin Deficiency Diseases* (Baltimore: The Williams & Wilkins Company, 1937), 163.

Flood), vitamin X had dwindled to the seriously deficient level which characterizes it today. Subsequent chapters will explain why vitamin X availability dwindled this way.

Conclusion

I am aware of nothing about "old age" which is inexplicable in terms of it being due to a nutritional deficiency disease resulting from lack of a previously unknown vitamin. This category of disease seems to provide a complete explanation of the facts in regard to "old age" and human longevity. To the best of my knowledge, this is true of no other category of known diseases.

Just as scurvy results from a diet deficient in vitamin C, so "old age" results from a diet deficient in vitamin X. Vitamin C is the anti-scurvy vitamin. Vitamin X is the anti-aging vitamin.

The Flood somehow broke the natural supply of vitamin X. This caused human life spans to be reduced ultimately by more than a factor of ten relative to what they had been prior to the Flood.

48 *Aging: Cause and Cure*

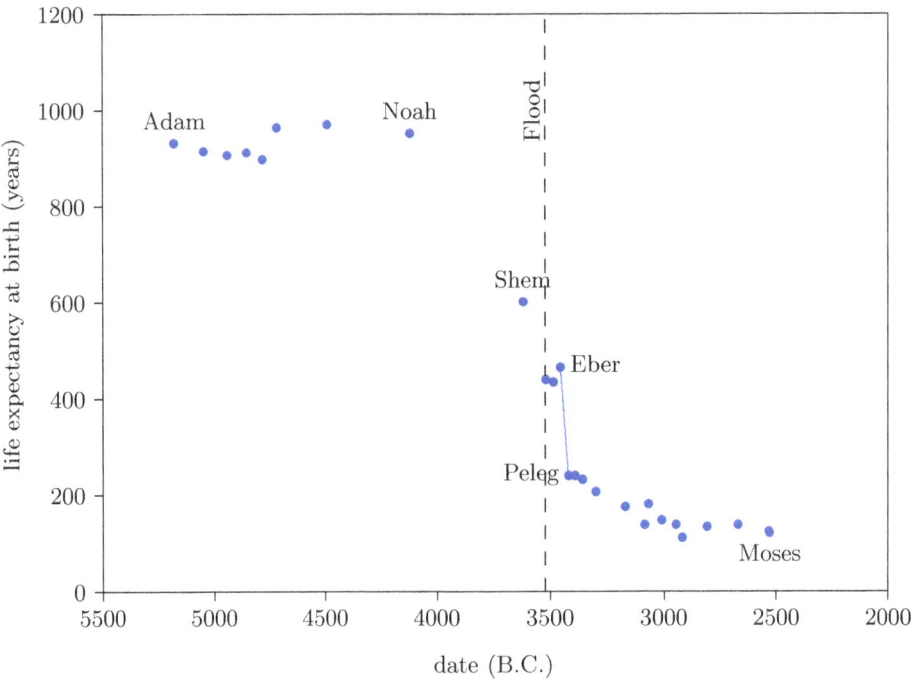

Figure 4: Biblical life expectancy at birth data for selected males.

Chapter Five

Properties of Vitamin X

Having come to understand that "aging" results from inadequate dietary vitamin X, the research goal becomes to identify the chemical compound—the molecule—corresponding to vitamin X. Once vitamin X has been identified, the expectation is that restoring it to human diets will cure "aging."

But identifying the compound corresponding to vitamin X is no easy task. There are nearly an infinite number of chemical compounds to choose from. It will clearly not do to try random guesses. Rather, it is necessary, like good detectives, to gather every bit of information we can about vitamin X, and then to use these clues to try to deduce the molecular identity of vitamin X. The current chapter begins this process by revealing two important properties of vitamin X.

The Eber–Peleg Drop

The biblical life expectancy data (Figure 4) display a fairly uniform, natural decline from Noah to Moses. The only real irregularity in this uniform decline, allowing for normal scatter in the data points, is a sudden drop in life spans between Eber and Peleg. This irregularity is highlighted in Figure 4 by a line connecting the Eber and Peleg data points.

This sudden drop has been used by some to argue for gaps in the Genesis chapters 5 and 11 genealogies from which many of these life expectancy data points have been taken. The argument, in this case, is that it is unlikely that life spans would have changed so dramatically in a single generation—from 464 years for Eber to just 239 years for Peleg—when they changed little in both the preceding two generations and in the following two generations.

In point of fact, there are no generations missing between Eber and

Peleg. The sudden drop in life spans between Eber and Peleg is real history. That is, it really was the case that Peleg was Eber's direct, first-generation son. That this is the case will become increasingly clear in subsequent chapters.

From a modern perspective, this yields a very curious result. It says that the son, Peleg, died of "old age" nearly two centuries before his father, Eber, died of "old age." Said another way, the father was physiologically yet in mid-life when his son died of "old age."

How is such a thing possible?

Such a thing is possible if and only if vitamin X has a long biological half-life.

Biological Half-life

Biological half-life is a measure of how long a substance tends to remain in the body before it is eliminated. The biological half-life can be broken down into half-lives for individual organs, like the heart, or the kidneys. Different organs can have very different half-lives. For example, it is possible for a compound to clear from the kidneys very rapidly, but to be retained in heart muscle for a long time. It would then have a short half-life in the kidneys and a long half-life in the heart.

The overall biological half-life of a substance in the human body can be measured by giving a person a small dose of that substance and measuring the dose remaining in the body (all organs and tissues) after an elapsed time. The biological half-life is the time it takes for just one half of the original dose to be remaining.

Sodium, which we get from common table salt, sodium chloride, is used in tissues throughout the body. It has a relatively short biological half-life of just 29 days. Calcium, on the other hand, tends to get tied up in bone, and has a relatively long biological half-life of 49 years.[18]

Biological Half-life of Vitamin X

Imagine two modern individuals, Bob and Tom, of the same age—say twenty years. If neither is given supplemental doses of vitamin X, then

---

[18] Joseph W. Kane and Morton M. Sternheim, *Physics*, (New York: John Wiley & Sons, 1980), Table 33.1, 577.

both will die of vitamin X deficiency disease ("old age") within a few decades of seventy-five years. We know this with fair certainty because billions of individuals have proven it since the Flood.

Now imagine that Bob is given supplemental doses of synthetic vitamin X so that he gets all the vitamin X his body needs from age twenty on, while Tom remains at a natural, present-day (deficient) level of vitamin X. Tom will still contract "old age" and die of it within a few decades of 75 years. But not so Bob. Bob will not contract "old age" at all, any more than a person receiving an adequate diet of vitamin C will contract scurvy.

We see by this simple example that two individuals of the same age, living at the same time, can experience very different rates of "aging" depending on their respective dose rates of vitamin X.

Now imagine that Bob is given synthetic vitamin X for one year only, after which he, like Tom, receives only a natural present-day level of vitamin X. What will be the result?

If vitamin X has a biological half-life measured in days, like sodium, then the benefit to Bob of his year-long supplemental dose of vitamin X will be merely to increase his life expectancy by one or two years. Extra vitamin X can benefit Bob only while he has it in his body, and he will have it in his body for only a short while after he stops taking it, if it has a short biological half-life. But if vitamin X has a relatively long biological half-life, say 49 years, like calcium, then Bob would be expected to outlive Tom by a century or more. The reason for this is that, in the case of a long biological half-life, vitamin X continues to be maintained at high levels in Bob's body long after supplemental dosing has stopped.

This is what happened in the case of Eber and Peleg. Since no means of artificially synthesizing vitamin X was available, Eber and Peleg were limited to only that dose of vitamin X which the environment naturally provided. Eber was born at a time when the amount of vitamin X in the environment was much higher than it is today. Sometime during the thirty-four years between the birth of Eber and the birth of Eber's son, Peleg, the amount of vitamin X in the environment declined dramatically. There is a very good reason why the amount of vitamin X in the environment declined this way, which will be discussed in a subsequent chapter, but, for now, notice merely that, as a result of this decline, the natural dose of vitamin X received by Peleg was always much less than that which his father Eber had initially received.

From Peleg's birth on, both Eber and Peleg were receiving the same, relatively low, natural dose of vitamin X from the environment. But even though they were both limited to the same natural dose rate of vitamin X from Peleg's birth on, Eber did not die of "old age" at the same time Peleg did. Eber lived on for several centuries after Peleg, his son, had died of "old age." Eber outlived his son Peleg by over two hundred years because Eber carried a higher level of vitamin X with him, in his body, long after vitamin X had declined dramatically in the natural environment.[19] This shows that the biological half-life of vitamin X must be relatively long—on the order of a century.

The long biological half-life of vitamin X is apparent in the Genesis life span data in many other instances than just Eber and Peleg. Notice, as a single additional example, that Shem, who accompanied his father Noah on the ark, outlived not only Peleg, his great-great-grandson, but also even Terah, Peleg's great-great-grandson (and father of Abraham). Shem, born before the Flood, died of "old age" only 25 years before Abraham—born 350 years after the Flood—died of "old age."

Clearly, vitamin X has a long biological half-life.

Environmental Half-life of Vitamin X

The environmental half-life is the time it takes for just one half of an original amount of a substance to be remaining in the environment after it has been added to the environment. As with biological half-life, the environmental half-life can be broken down into various compartments. For example, the environment might be broken up into atmosphere, hydrosphere, and biosphere compartments. Each of these compartments would have its own characteristic half-life for a given compound, and these half-lives could differ substantially from one another.

Vitamin X displays two environmental half-lives, a short one and a long one, implying its presence in two environmental compartments. The long half-life compartment is apparent from the slow decline in life expectancies between Peleg and Moses (Figure 4, page 48). This slow de-

---

[19] It does not need to be vitamin X itself which is carried forward in time in the body. One or more metabolites of vitamin X will also do. I am ignoring the distinction between vitamin X and its metabolites in an effort to keep the present discussion as simple as possible.

cline compartment will be discussed further in subsequent chapters. The short half-life compartment is apparent from the rapidity of the drop in life expectancy between Eber and Peleg. This drop in life expectancy requires that the environmental abundance of vitamin X dropped dramatically between Eber's birth and Peleg's birth. The time between these two births was just 34 years. Thus we learn that vitamin X in the natural environment can decline dramatically in a matter of a few decades or less. This is just another way of saying that vitamin X has an environmental compartment with a relatively short half-life—on the order of a decade or less.

Conclusion

Vitamin X has an unusually long biological half-life. It has at least two environmental compartments, one with a relatively short half-life and the other with a considerably longer half-life.

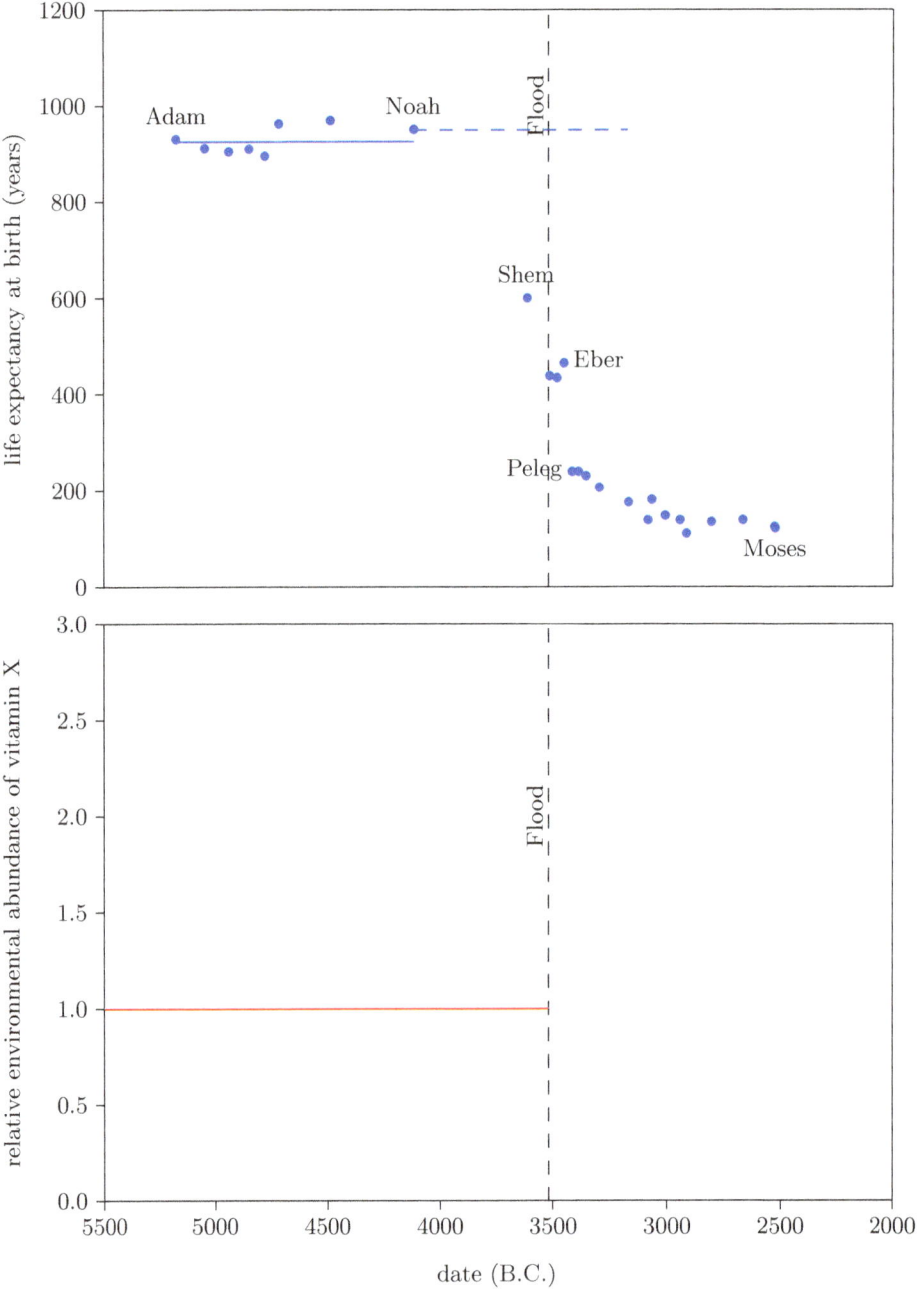

Figure 5: Top: Biblical life expectancy at birth data for selected males. Bottom: environmental abundance of vitamin X relative to the pre-Flood.

CHAPTER SIX

THE ENVIRONMENTAL ABUNDANCE OF VITAMIN X

We are now able to begin to piece together how the environmental abundance of vitamin X changed as a result of the Flood. By focusing on the Eber–Peleg drop in life expectancies, we have already learned that the environmental abundance of vitamin X declined dramatically between Eber and Peleg. Noah's life span reveals another important change in the environmental abundance of vitamin X which was brought about by the Flood.

Noah's Life Span

Noah's life span shows that the environmental abundance of vitamin X was initially caused to *increase* by the Flood. This may be seen as follows.

The dashed blue line in the top graph of Figure 5 represents Noah's life from birth until death. Notice that Noah lived the final centuries of his life in the post-Flood, post-Peleg regime, when the environmental abundance of vitamin X had dropped to low levels compared to the pre-Flood level. Noah lived his first 600 years before the Flood, and his final 350 years after the Flood. Roughly 250 years of Noah's life were lived in the post-Peleg regime. This seems to imply that Noah's life span should have been *shorter* than the average pre-Flood life span, when, in fact, it was *longer*.

The average life span of the seven pre-Flood men preceding Noah in the biblical life expectancy data set is 926 years. This average is shown in the top graph of Figure 5 by the solid, blue line extending from Adam to Noah. Notice that Noah's life expectancy at birth data point lies above this line. This shows that Noah's life span was *longer* than average for a pre-Flood male.

How is this possible, given that Noah lived the final two and a half centuries of his life in an environment which was seriously deficient in vitamin X?

The biological half-life of vitamin X is not a sufficient answer to this question. From the birth of Peleg on, life expectancies were dramatically shorter than they had been prior to the Flood. As just mentioned, Noah spent the last 250 years of his life in this post-Peleg regime. Even if the biological half-life of vitamin X were 200 years (and it is not that large, as will be shown in Chapter 10), by the time Noah died, his body concentration of vitamin X would have dropped by more than a factor of two relative to his pre-Flood body concentration. Thus his average body concentration of vitamin X during these post-Peleg years would have been less than 75% of what it had been pre-Flood. Said simply, even given a long biological half-life for vitamin X, Noah still would have suffered serious vitamin X deficiency during the post-Peleg years (though not nearly as serious as Peleg's vitamin X deficiency), and this should have shortened his post-Flood life span significantly. Notice that Noah's son, Shem, had his life span reduced from the pre-Flood average of 926 years to just 600 years as a result of living roughly 400 years in the post-Peleg regime. This yields a loss of 326 years of average pre-Flood life span per roughly 400 post-Peleg years. At the same rate, Noah should have lost roughly 200 years relative to the pre-Flood average life span. Instead, Noah lived 24 years *longer* than the average pre-Flood life span. Clearly, the long biological half-life of vitamin X is not sufficient to explain Noah's longer-than-average life span.

Noah's above average life span requires that Noah entered the post-Peleg years with a concentration of vitamin X in his body which was higher than the pre-Flood concentration. This higher concentration could have been obtained only after the Flood and before Peleg. Thus, the Flood must have caused the environmental abundance of vitamin X initially to increase.

The Environmental Abundance Graph

The bottom graph of Figure 6 shows both the increase caused by the Flood (vertical arrow pointing up), and the sudden decline between Eber and Peleg (vertical arrow pointing down). The slope of the decline between Eber and Peleg is unknown, other than that it was steep. Keeping things simple, I have assumed that the drop was instantaneous, as the arrow depicts. The arrow is plotted midway between the births of Eber and Peleg.

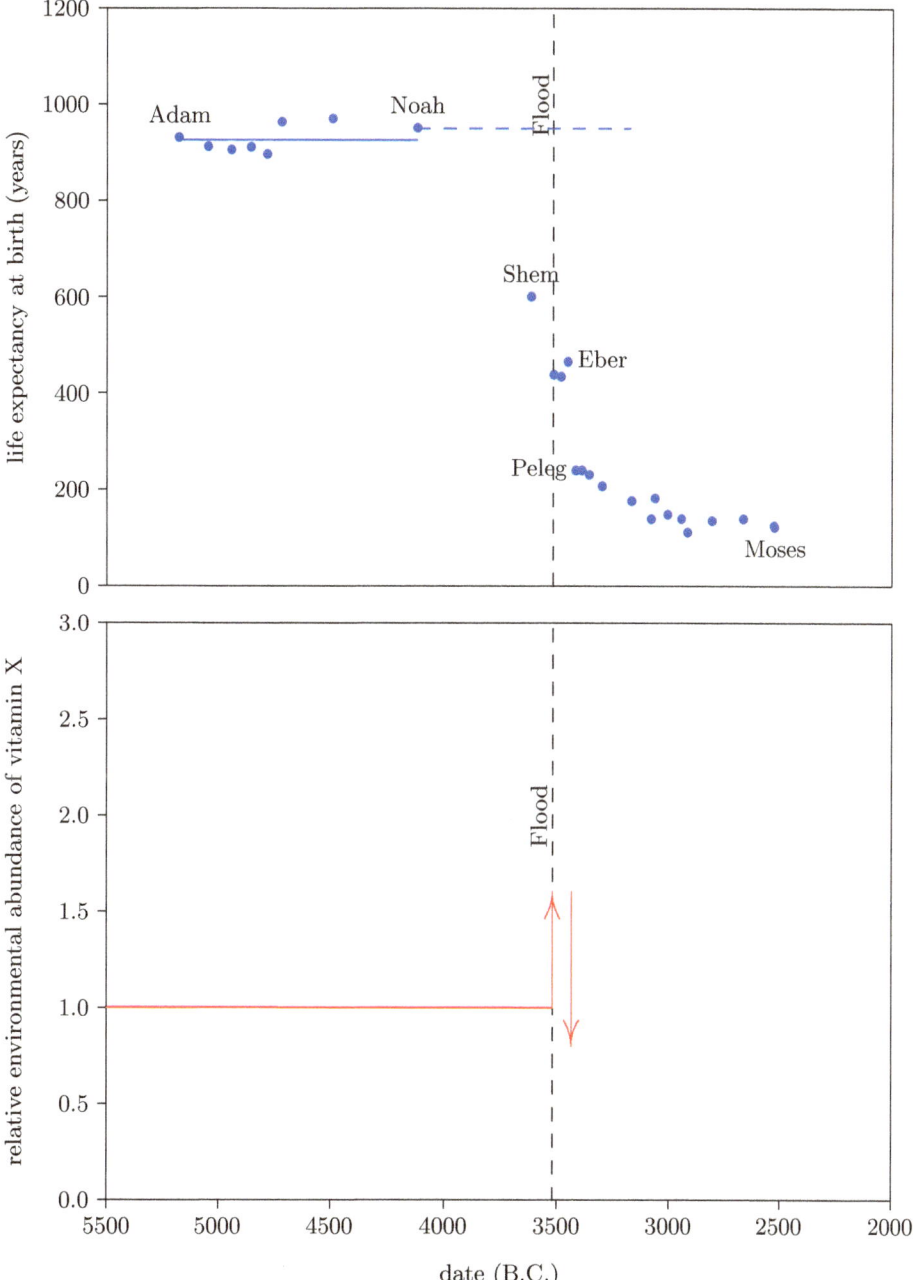

Figure 6: Top: Biblical life expectancy at birth data for selected males. Bottom: environmental abundance of vitamin X relative to the pre-Flood.

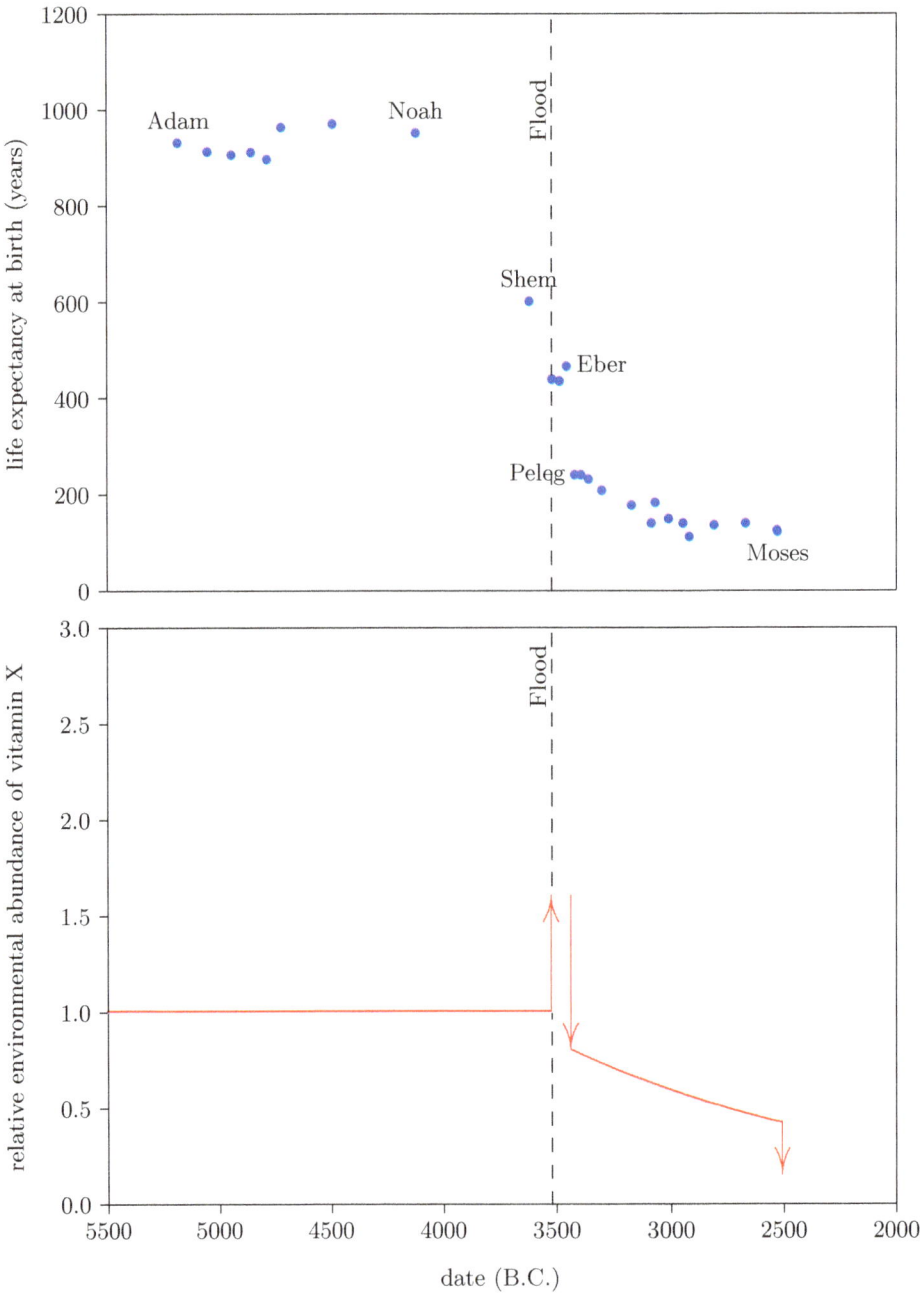

Figure 7: Top: Biblical life expectancy at birth data for selected males. Bottom: environmental abundance of vitamin X relative to the pre-Flood.

The Post-Peleg Decline

From Peleg to Moses, life expectancies show a relatively gentle decline (Figure 7, top graph). This decline in life spans implies an underlying decline in the environmental availability of vitamin X, which I have illustrated in the bottom graph of Figure 7.

The decline in life spans is fastest initially, so I have chosen to represent the decline by a natural exponential decay. At this point, we don't know the initial height of the exponential, and we don't know how fast it should drop off. I have made initial guesses at these two parameters to get the representative exponential decline which is plotted in the bottom graph of Figure 7. These guesses will be mathematically refined in a subsequent chapter.

Notice that this adds a second environmental half-life to vitamin X, one which is quite long, as mentioned previously. This says that vitamin X is present in two environmental compartments.

The Moses Drop

Life spans appear to have rapidly dropped once again in the lifetime of Moses. In the first half of the tenth verse of Psalm 90, Moses notes that life spans had dwindled to roughly 75 years.

> As for the days of our life, they contain seventy years,
> Or if due to strength, eighty years,...

Moses himself lived 120 years, and his brother Aaron lived 123 years. Thus, there was another Eber–Peleg type of drop in life spans during Moses' lifetime. This drop appears to have resulted in the modern life span for humans, which has persisted now for more than 4500 years. This final drop is indicated by the final red arrow. For simplicity, I have chosen to make this drop instantaneous once again. In real life, this final drop would almost certainly have been more gradual.

A date for this drop can be estimated as follows. Because Moses had to observe and record the drop, the latest time that persons could have begun dying of "old age" at 75 years on average is the same time Moses himself died of "old age" at 120 years. These persons would have been born when Moses was 45 years old. This places the latest date for the drop in 2482 B.C. The earliest time this drop could have happened, and Moses and Aaron still enjoy their recorded, significantly longer life spans due to the biological half-life of vitamin X, is the year following the birth of Moses. This is 2526 B.C. The best estimate, in this case, is the midpoint between earliest and latest possibilities, which is 2504 B.C.

We do not know, at present, how far the environmental abundance of vitamin X dropped at this point. This will clarify in subsequent chapters.

Conclusion

The environmental abundance of vitamin X breaks naturally into four distinct time periods. These are depicted in the bottom graph of Figure 8.

The first is the "Pre-Flood" time period. This is a steady state region, in which the environmental availability of vitamin X was constant and relatively high, though still somewhat short of dietary sufficiency. This region ended with the coming of the Flood.

The second is the "Spike" time period, initiated by the Flood. This was a relatively brief period during which the environmental abundance of vitamin X was higher than in any other time period. The Spike ended with a sharp reduction in environmental availability of vitamin X sometime between the births of Eber and Peleg.

The third is the "Decline" time period. Environmental availability of vitamin X, which was already seriously deficient at the beginning of this time period, slowly declined, becoming ever more seriously deficient during this time period.

The fourth is the "Modern" time period. It was ushered in roughly 2504 B.C. by a final rapid drop in life spans to the modern value near 75 years, which took place during the lifetime of Moses.

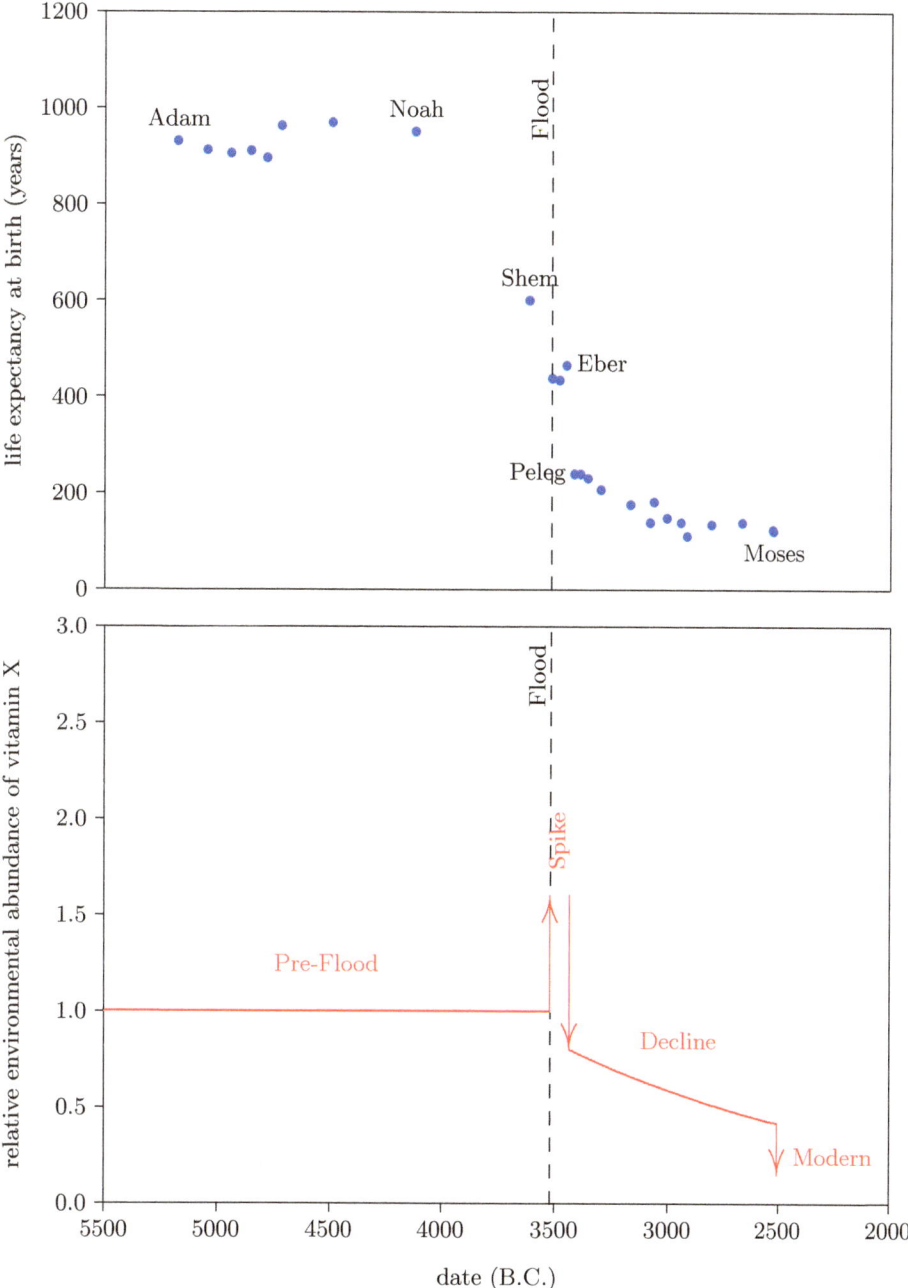

Figure 8: Top: Biblical life expectancy at birth data for selected males. Bottom: environmental abundance of vitamin X relative to the pre-Flood.

CHAPTER SEVEN

THE NATURAL SYNTHESIS AND DISTRIBUTION OF VITAMIN X

It is now possible to advance reasonable answers to three key questions.

1. How was vitamin X distributed by nature?

2. How was vitamin X synthesized by nature?

3. How did vitamin X enter the human diet?

The answers to these questions provide additional clues to the identity of vitamin X.

How was Vitamin X Distributed by Nature?

The biblical longevity data seem to imply that, while longevity varies widely in different periods of history, it is globally uniform at a given point in time. The biblical longevity data do not encompass a global population, of course, but they do extend over a large enough geographical area in the Middle East—including at least portions of modern Iraq, Turkey, Israel and Egypt—to make global uniformity seem likely.

Is there any way of explaining a point-in-time geographical uniformity of vitamin X given its large temporal variations?

Yes. The atmosphere is present globally, and it mixes rapidly relative to human life spans—just a year or two from pole to pole. If vitamin X were distributed by the atmosphere, then its consequences for longevity would be globally uniform at any point in time.

The atmosphere is made up of gases. The mixture of gases comprising the atmosphere is called air. Nitrogen and oxygen are the main gases in air, but there are many gases present in minor amounts as well. For example, nearly one percent of air is the gas called argon. Well known gases in air include ozone, carbon dioxide, and methane. Dozens more, not so well known, like carbon disulfide and nitrogen dioxide, might easily be mentioned. All of the gases in air, other than nitrogen and oxygen, are called trace gases. Point-in-time global uniformity would be explained if vitamin X were a trace gas or were somehow involved with a trace gas.

The idea that vitamin X is somehow involved with a trace gas provides an explanation of the Decline time period of the vitamin X environmental abundance graph. Notice that the Decline lasted roughly 900 years (Figure 8, page 61). The Flood seems to have broken the fundamental source of vitamin X, but vitamin X did not drop immediately to zero following the Flood. Rather, following an initial spike, it slowly declined for 900 years. Involvement of a trace gas provides an obvious explanation of this prolonged decline. Specifically, there is a component of earth's global environmental system which has this characteristic timescale. One thousand years is the order-of-magnitude timescale for the turnover of earth's oceans. Deep ocean water takes roughly a thousand years to move from the bottom of the ocean to the surface. A change lasting 900 years would be expected if the oceans acted as a reservoir for the conjectured atmospheric trace gas. Gases have a water solubility. Gas molecules, present in earth's oceans, are effectively cut off from the atmosphere. They can get out of the ocean and into the atmosphere only by diffusion through the ocean–atmosphere interface at the surface of the ocean. Gas molecules down deep in the ocean will be stuck in the ocean for hundreds of years waiting for the water in which they are trapped to move slowly up to the surface, where they can vent to the atmosphere via diffusion through the surface. The 900 years of the Decline time period can be readily explained as the time it took for the trace gas involved with vitamin X to empty out of the oceans into the atmosphere.

Said succinctly, though the Flood broke (i.e., shut down) the primary source of some atmospheric trace gas, the oceans continued to act as a source of this gas to the atmosphere until their supply had been exhausted roughly 900 years later.

All of this argues forcefully for the idea that vitamin X was distributed globally via the atmosphere, either as a trace gas or as some atmospheric product of a trace gas.

Notice that this also explains why the environmental abundance of vitamin X has a fast environmental half-life compartment and a slow environmental half-life compartment. The fast compartment is the atmosphere, and the slow compartment is the oceans.

How was Vitamin X Synthesized by Nature?

The idea that vitamin X is a trace gas, while possible, is less attractive than the idea that vitamin X is a product of a trace gas. The reason for this has to do with the nature of the thirteen traditional vitamins and the nature of the atmospheric chemistry of trace gases.

To begin with, of the thirteen traditional vitamins, none are gases. Vitamin-likeness argues against vitamin X being a gas.

The atmosphere contains a lot of oxygen. Atmospheric chemistry tends either to break a trace gas down into smaller fragments, or to add a few oxygen atoms to a trace gas. Breaking a trace gas down into smaller fragments will yield yet lower molecular weight products, which will tend to be gases. Thus, vitamin-likeness favors the oxygen-addition process over the fragmentation process.

Atmospheric chemistry argues for vitamin X being a small molecule. Atmospheric trace gases tend to be small molecules. Large molecules are too massive to have sufficient vapor pressure to be present in the gas phase. There is no sharp mass cutoff for presence of a compound in the atmosphere, but a molecular weight of 250 g/mole seems to be a reasonable dividing line above which few molecular substances are likely to be found in significant abundance in the atmosphere.

This small-molecule dividing line places vitamin X in the water-soluble vitamin category. Of the thirteen traditional vitamins, nine are water-soluble and four are fat-soluble. None of the fat-soluble vitamins has a molecular weight below 250 g/mole, while five of the water-soluble vitamins have molecular weights below this cutoff. Thus, vitamin-likeness argues for vitamin X to be a small, water-soluble molecule.

Does the atmospheric chemistry, oxygen-addition process tend to produce water-soluble products?

Yes. Oxidation of trace gases tends to yield small acids as stable end products. Small acids are notoriously water soluble.

Are any of the small water-soluble vitamins acids?

Yes. In fact, of the five water-soluble vitamins having molecular weights below 250 g/mole, four are acids.

Thus, the idea that vitamin X is a small acid, an oxidation product of an atmospheric trace gas, is very attractive.

How did Vitamin X Enter the Human Diet?

This idea immediately answers the third question. Because small acids are soluble in water, acids produced in the atmosphere are rapidly washed out of the atmosphere by rain once they have been formed. This means that they are naturally present in rainwater, in freshwater ponds and lakes, and in rivers. Clearly, vitamin X would have entered human diets in antiquity via the water that humans drank.

The MSA Example

Methanesulfonic acid (MSA) is an example of a small organic acid which is produced in the atmosphere by oxidation of a trace gas.[20] The trace gas in this instance is the extensively studied DMS (dimethyl sulfide).

DMS is naturally produced in ocean surface waters from DMSP (dimethylsulfoniopropionate), which is a metabolite of some marine algae. In the atmosphere, DMS reacts with the hydroxyl radical (OH·) or the nitrate radical (NO_3·) resulting in oxidation to a variety of products, including MSA.

MSA concentrations in glacier ice are generally around 10 micrograms per liter. Thus, use of rainwater for drinking today (not generally recommended due to ubiquitous pollutants in modern air) would naturally supply human diets with 20 micrograms or more of MSA per day. This is not to suggest that MSA is desirable in human diets today, but rather to show that nature is capable of producing and supplying to the human diet compounds in vitamin-like quantities via an atmospheric trace gas.

Conclusion

The idea that vitamin X is a small acid, an oxidation product of an atmospheric trace gas, is compelling. It provides satisfying explanations of the natural synthesis, distribution, and dietary presence of vitamin X.

---

[20]John H. Seinfeld and Spyros N. Pandis, *Atmospheric Chemistry and Physics* (New York: John Wiley & Sons, Inc., 1998), 315–317.

Chapter Eight

What the Flood Broke

Having deduced that vitamin X seems likely to be an oxidation product of an atmospheric trace gas, the next strategic goal is to try to identify the trace gas, as an aid to identifying vitamin X. There are, unfortunately, still way too many trace gases (giving rise to a plethora of oxidation products) to make random guessing a feasible research strategy. It is necessary to make use of every clue once again. But this time, it is knowledge of the nature of the Flood, not the properties of the trace gas, which yields the vital breakthrough.

The Nature of the Flood

We are looking for that particular trace gas which is the precursor to vitamin X. This precursor gas was being supplied to the atmosphere during the Pre-Flood in much greater abundance than it is at present. The Flood somehow broke the supply to the atmosphere of this precursor gas.

How might the Flood have broken the supply of an atmospheric trace gas? The answer to this question seems likely to provide the all-important clue to learning the identity of vitamin X. Thus it appears that the cure to human "aging" requires "only" a thoroughgoing understanding of the true nature of Noah's Flood. Fortunately, I have such an understanding.

Let me reword that. In actual fact, good fortune has had little to do with it. I have spent decades of my life uncovering the truth regarding Noah's Flood. I have found that mainstream academia is seriously mistaken about the Flood. Modern academicians seem to regard it as essentially mythological, if they bother to think about it at all. And I have found that individuals holding out against the view that the Flood was a myth (usually for theological reasons) are generally seriously mistaken in their understanding of the true nature of the Flood. In consequence, none are very well positioned to be of much help in regard to the present question. It was necessary for me to solve the mystery of the true nature of Noah's Flood before I could hope to solve the mystery of "aging."

The important thing to know about the Flood in the present context is that it was hemispherical in extent. It resulted from the waters of the southern hemisphere flowing up into the northern hemisphere and heaping up to great depth there for some months. Why this happened—the geophysics behind it—is explained in detail in my book, "Noah's Flood Happened 3520 B.C.,"[21] so I will not go into detail here. The important thing to notice at present is only this: heaping the planet's oceans up in the northern hemisphere will expose the sea floor in the southern hemisphere. In fact, the sea floor in the high southern latitudes surrounding Antarctica was entirely uncovered for at least 110 days during the Flood.

The Nature of Sea Floors

Sea floors are hidden from our eyes. As a result, they are not items of common knowledge. The important thing to know about them at present is that they are covered with sediments. The average depth of the sediments covering the floors of the world's oceans exceeds a quarter of a mile.

Sea floor sediments are cut off from atmospheric oxygen. In consequence, they are anaerobic. Anaerobic microorganisms in organic-rich sediments produce organic gases as byproducts within those sediments. Methane is a major, well known byproduct gas of anaerobic digestion. Other anaerobic byproduct gases include, for example, carbon dioxide, hydrogen, nitrogen and hydrogen sulfide.

The Source of the Precursor

I suggest that the source of the atmospheric trace gas which is the precursor to vitamin X was anaerobic digestion of organic detritus within the sea floor sediments of the southern oceans. Steady state diffusion of molecules of this trace gas from the sediments to the overlying ocean

---

[21]Gerald E. Aardsma, *Noah's Flood Happened 3520 B.C.* (Loda, IL: Aardsma Research and Publishing, 2015). www.BiblicalChronologist.org.

water is what the Flood broke. The Flood broke this steady state diffusion by removing the great depth of water which normally covers the sediments. This reduced the pressure these sediments are normally under by hundreds of atmospheres. (Each ten meters of water depth adds another atmosphere of pressure, and the average depth of the oceans is 3800 meters.) The inevitable result was rapid release (outgassing) to the atmosphere of the anaerobic byproduct gases contained within these sediments.

Conclusion

The Flood broke the steady state diffusion of anaerobic gases from the sediments to the bottom waters of the southern oceans by emptying southern sediments of their gases during the year of the Flood.

This explains why the Flood was followed by the Spike. The Flood released thousands of years' worth of vitamin X precursor gas from the sediments into the atmosphere all in one year. The result was decades of abnormally high vitamin X production in the atmosphere during the Spike.

When the atmosphere had finally managed to cleanse itself of vitamin X precursor (and the load of other sea floor gases), only precursor diffusing out of the water of the oceans remained to supply vitamin X. This explains the Decline.

When the oceans' supply of precursor ran out, life spans dropped to near 75 years, initiating the Modern time period.

The sediments of the southern oceans were emptied of their store of precursor gas by the Flood. This vital source of precursor has so far not recovered in five and a half thousand years.

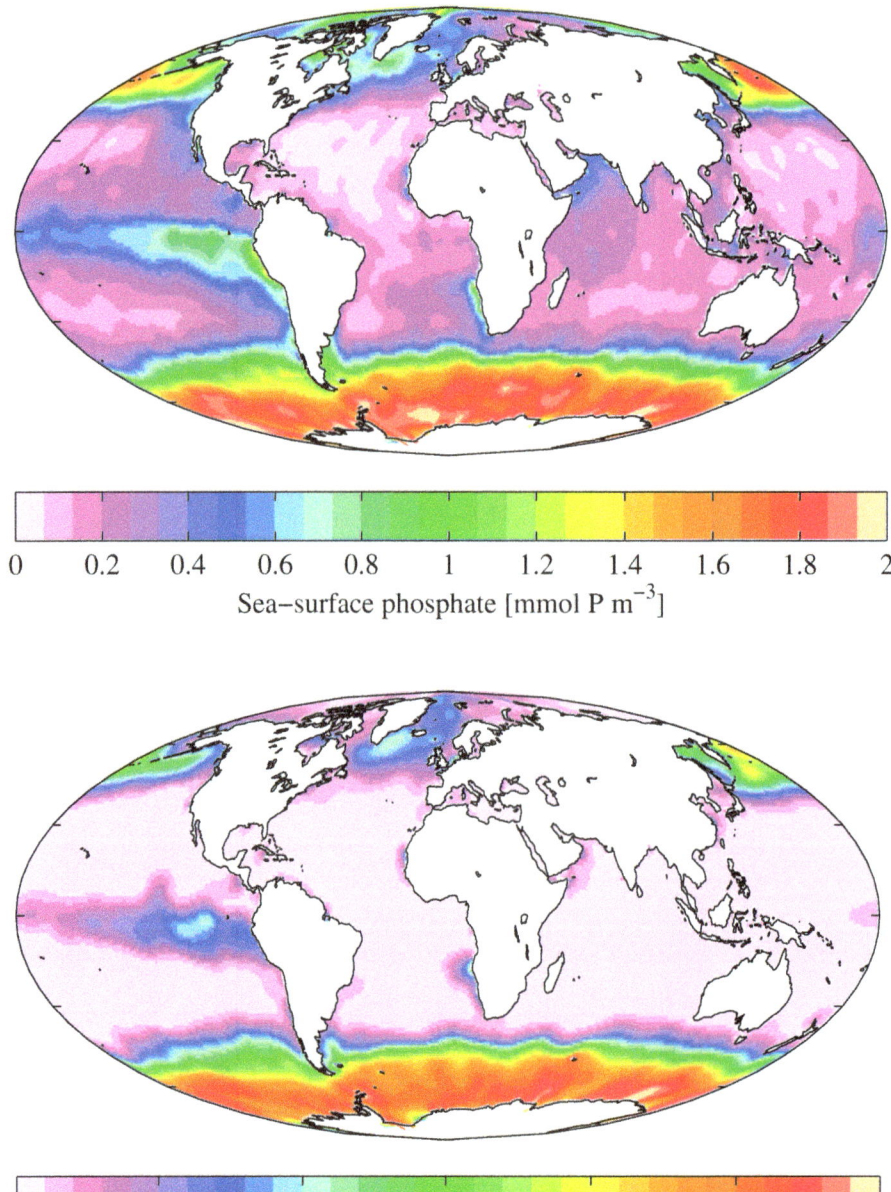

Figure 9: 2009 annual phosphate and 2009 annual nitrate at the surface. https://en.wikipedia.org/wiki/World_Ocean_Atlas [Images Attribution: By Plumbago (Own work) [CC BY-SA 3.0 (http://creativecommons.org/licenses/by-sa/3.0)], via Wikimedia Commons]

Chapter Nine

Vitamin X Revealed

It seems clear that the vitamin X precursor gas was predominantly produced in the sediments of the southern oceans. If its production had been spread out over the sea floor of the entire planet, then the supply of precursor gas would have been reduced only roughly in half by the Flood, and it seems unlikely that reduction of the environmental abundance of vitamin X by a factor of two could have resulted in the observed reduction of life spans by more than a factor of ten. Furthermore, the supply of precursor gas should then have shown some recovery over the past five and a half thousand years, with a correspondingly noticeable recovery in life spans, due to a natural recovery of precursor gas within those sediments which had been only slightly depressurized by the Flood. But no such recovery in life spans has been seen. Thus, available data seem to favor the idea that nearly the entire source of precursor was destroyed by the Flood.

This focuses attention on the southern sea floor. What is so special about it? Why did depressurization of just the southern sea floor destroy nearly the entire source of vitamin X precursor gas?

Figure 9 supplies answers to these questions. It shows that the surface waters of the oceans surrounding Antarctica are especially rich in both phosphate and nitrate. Phosphate and nitrate are broadly important and often limiting nutrients within the marine environment. The existence of both together in high concentrations in the ocean surface waters surrounding Antarctica more or less guarantees accumulation of organic-rich sediments on the sea floor surrounding Antarctica.[22] Indeed, it seems appropriate (if a trifle prosaic) to regard the sea floor surrounding Antarctica as the planet's septic tank.

---

[22]Sigman, D. M. and Hain, M. P. "The Biological Productivity of the Ocean," *Nature Education Knowledge* 3.10 (2012) 21.

A Phosphorus Trace Gas

The concentration of phosphorus in surface waters surrounding Antarctica makes the planet's septic tank a rich depository of phosphorus-laden biomolecules. The presence of this rich accumulation of biophosphorus instigates the hypothesis that the vitamin X precursor was a phosphorus gas byproduct of anaerobic digestion within pre-Flood sea floor sediments.

This hypothesis seems to encounter an immediate difficulty. Phosphorus is well-known not to have any significant trace gases. In fact, phosphorus contrasts with other elemental environmental cycles in not having any significant atmospheric component. For example, the nitrogen cycle has an obvious atmospheric component in the form of N_2 gas. Similarly, the oxygen cycle has an obvious atmospheric component in the form of O_2 gas. The sulfur cycle has the already-mentioned DMS as a significant atmospheric component. DMS moves 15–25 million metric tons of sulfur from the oceans into the atmosphere each year.[23] The environmental iodine cycle has methyl iodide (MeI) as an atmospheric component. The environmental carbon cycle has carbon dioxide (CO_2) as an atmospheric component. ... But there is no atmospheric component in the case of phosphorus.

I suggest that this difficulty is apparent only. If one supposes that the phosphorus cycle is presently operating in steady state—that it has been behaving for time out of mind just as it is observed to be behaving at present—then the presently observed lack of an atmospheric phosphorus component leads immediately to the conclusion that it is the nature of the phosphorus cycle to be without any significant atmospheric component. But as soon as one knows the truth about the Flood, one knows better than to suppose that the phosphorus cycle is presently operating in steady state. I suggest, in fact, that phosphorus normally has an atmospheric component, in analogy with other elemental biogeochemical cycles, but that the present time is not normal, at least as far as phosphorus is concerned. I suggest that the atmospheric component of the environmental phosphorus cycle is the fundamental thing which the Flood broke. And I suggest that it was this breakage which resulted in the loss of human longevity following the Flood—that loss of the atmospheric component

---

[23]John H. Seinfeld and Spyros N. Pandis, *Atmospheric Chemistry and Physics* (New York: John Wiley & Sons, Inc., 1998), 59.

of the phosphorus cycle (i.e., loss of this phosphorus trace gas) meant loss of the vitamin X precursor, resulting in loss of vitamin X itself.

The fact that the environmental phosphorus cycle is not represented by any atmospheric component *today* does not mean that this was the case before the Flood. Phosphorus gases do exist and are well known in the chemistry laboratory and in industry.

Phosphine (PH_3) is the simplest gas containing phosphorus. It is a possible pre-Flood phosphorus trace gas candidate, but it is not a possible vitamin X precursor candidate. Vitamins are organic compounds (i.e., they contain one or more carbon atoms). Phosphine is not organic. Atmospheric chemistry can add oxygen atoms to a precursor gas; it cannot add carbon atoms. Thus, the precursor gas must start out as an organic compound.

Simple organic trace gases of the elements often result from addition of one or more methyl groups (CH_3) to the element. An example is methyl iodide (MeI), already mentioned. MeI is sourced to the atmosphere from the oceans. The oceans also source methyl chloride (MeCl) and methyl bromide (MeBr) to the atmosphere. Dimethyl sulfide (DMS or Me-S-Me) is an example of a trace gas containing two methyl groups. It, too, is sourced to the atmosphere from the oceans.

We are looking for a simple organophosphorus trace gas sourced to the atmosphere from the oceans. Methylated phosphorus gases seem to be the obvious choice.

Methylated Phosphorus Gases

The biological pathway by which methylated phosphorus gases might arise in anaerobic sedimentary environments has yet to be demonstrated by science. It seems, however, that discovery of such a pathway is imminent.

> There are strong indications that phosphine and other reduced phosphorus compounds can be produced biogenically.[24]

Two different modes of production of methylated phosphorus gases can be imagined: 1. biomethylation of phosphorus, and 2. catabolism of phosphonates.

---

[24]Joris Roels and Willy Verstraete, "Biological formation of volatile phosphorus compounds," *Bioresource Technology* 79 (2001) 243–250.

Biomethylation in general is well studied and reasonably well understood.

> Biomethylation is the process whereby living organisms produce a direct linkage of a methyl group to a metal or a metalloid, thus forming metal-carbon bonds. Methylation has been extensively studied and biomethylation activity has been found in the soil, but mainly occurs in sediments in e.g. estuaries, harbors, rivers, lakes and oceans. ... Anaerobic bacteria are believed to be the main agents of biomethylation in sediments and other anoxic environments.[25]

From a purely utilitarian perspective, however, the second possibility, catabolism of phosphonates, seems to me to be somewhat more attractive than biomethylation.

Phosphonates are chemical compounds having a carbon atom (C) bonded to a phosphorus atom (P). Catabolism is the process by which living organisms break down complex molecules into simpler molecules during metabolism.

In the pre-Flood anaerobic, organic-rich, phosphate-rich marine sedimentary environment surrounding Antarctica, microbes might naturally be expected to have been present which possessed an ability to "mine" phosphonates for their oxygen atoms while discarding the less readily catabolized C-P moieties as MeP "tailings," since both carbon and phosphorus would be abundantly available in more easily catabolized forms. The great versatility of microbial life makes it seem likely that some anaerobic microbe having ability to exploit this environment in this way should exist.

From chemistry, it is known that phosphorus can add one, two, or three methyl groups, giving methylphosphine (MeP), dimethylphosphine (DMeP), and trimethylphosphine (TMeP), respectively. Of these three, TMeP is the least interesting for the present purpose. It has a boiling point of 38–40°C, which means that, in the laboratory, it is a liquid rather than a gas. TMeP vapor oxidizes in the atmosphere to trimethylphosphine oxide (TMePO). TMePO is not easily oxidized a second time, and it is a severely hygroscopic solid, which means that it will be rapidly

---

[25] P. J. Craig and R. O. Jenkins, "Organometallic compounds in the environment: An overview," *Organic Metal and Metalloid Species in the Environment*, ed. Alfred V. Hirner and Hendrik Emons (New York: Springer, 2004), 5.

washed out of the atmosphere. If vitamin X is a small acid, as expected, then TMePO, which is not an acid, is not vitamin X.

Both DMeP and MeP are true laboratory gases. Both yield small acids upon oxidation in the atmosphere.

DMeP oxidizes first to dimethylphosphine oxide (DMePO). This is a liquid with a boiling point of 220°C. Like TMePO, this is likely to wash out of the atmosphere. However, it can oxidize further to dimethylphosphinic acid (DMePiA).

The final methylated phosphorus compound and potential vitamin X precursor, the gas MeP, oxidizes first to methylphosphine oxide (MePO). MePO is not very stable. It easily oxidizes a second time to methylphosphinic acid (MePiA). This small acid is hygroscopic; it will wash out of the atmosphere. However, it is readily susceptible to further oxidation, either in the gas phase or in aqueous solution, yielding methylphosphonic acid (MePA) as the stable final product.

Thus, two precursor gases (DMeP and MeP) and three vitamin X candidates (DMePiA, MePiA, and MePA) result from methylated phosphorus.

Examining the Methylated Phosphorus Vitamin X Candidates

MePiA is the least interesting of the three vitamin X candidates. Its susceptibility to oxidation implies that it will likely not survive long enough to get out of the atmosphere and into humans. In addition, I have found that this substance exhibits chronic toxicity when included in the diets of laboratory mice.

DMePiA is also of low interest. DMePiA is a phosphinate, having two carbon atoms bonded to the phosphorus atom. This arrangement (two carbons bonded to phosphorus) appears to have a severely limited natural biological utility relative to the phosphonates (which have just one carbon atom bonded to the phosphorus atom, as with MePA).

> Phosphinothricin (PT), a non-proteinogenic amino acid found in a number of peptide antibiotics, is the only known phosphinic acid natural product.[26]

---

[26] William W. Metcalf and Wilfred A. van der Donk, "Biosynthesis of Phosphonic and Phosphinic Acid Natural Products," *Annual Review of Biochemistry* 78 (2009) 65–94.

In sharp contrast, phosphonates appear to be ubiquitous, with especially large abundance in the marine environment[27] and some marine organisms.

> ... eggs of the freshwater snail Helisoma contain 95% of their phosphorus in the form of 2-aminoethylphosphonate-modified phosphonoglycans, whereas the sea amenaea Tealia possesses up to 50% of its phosphorus in a variety of phosphonolipids, phosphonoglycans and phosphonoglycoproteins. Other organisms, such as the protist Tetrahymena, have lower overall levels of phosphonate, but still synthesize as much as 30% of their membrane lipids in the form of phosphonolipids. The prevalence of C-P compounds in nature is perhaps best exemplified by the recent discovery that as much as 20-30% of the available phosphorus in the world's oceans is comprised of phosphonic acids.[28]

Notice that this also implies that a catabolic route to vitamin X precursor formation exists only for MeP. The catabolic production of DMeP would require abundant phosphinate substrate, which, as we have just seen, does not appear to exist in nature.

Thus the lack of biological utility of the phosphinates and the lack of a catabolic route to formation of DMeP as vitamin X precursor render DMePiA of low interest.

This leaves MePA as the sole interesting vitamin X candidate arising from methylated phosphorus.

Conclusion

I propose that MePA is, in fact, vitamin X, and that MeP is the vitamin X precursor gas (Figure 10).

MePA is the simplest member of the phosphonate class of compounds. Phosphonates as a class are biologically versatile compounds with "potent

---

[27] Xiaomin Yu et al., "Diversity and abundance of phosphonate biosynthetic genes in nature," *Proceedings of the National Academy of Sciences of the United States of America* 110.51 (December 17, 2013): 20759–20764.

[28] William W. Metcalf and Wilfred A. van der Donk, "Biosynthesis of Phosphonic and Phosphinic Acid Natural Products," *Annual Review of Biochemistry* 78 (2009) 65–94.

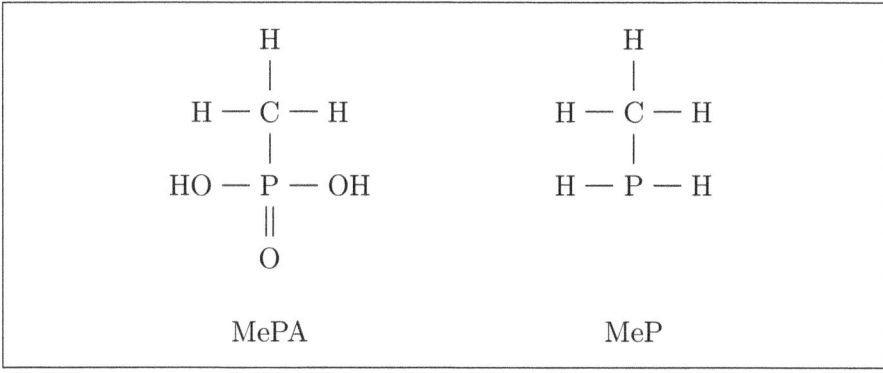

Figure 10: Methylphosphonic acid (MePA) and its atmospheric precursor gas methylphosphine (MeP).

bioactive properties."[29] MePA thus appears appropriately positioned to fill the role of a vitamin having anti-aging potency.

---

[29]William W. Metcalf and Wilfred A. van der Donk, "Biosynthesis of Phosphonic and Phosphinic Acid Natural Products," *Annual Review of Biochemistry* 78 (2009) 65–94.

Chapter Ten

Modeling the Biblical Life Expectancy Data

Having developed a qualitative explanation (a theory) of why human life spans plummeted following the Flood, my next task was to apply this theory quantitatively to the biblical life expectancy data. This was done by first putting the theory into mathematical form. To solve the resulting equations required use of numerical methods on a computer. I then applied the resulting mathematical computer model to the biblical life expectancy data.

It was necessary to model the data quantitatively for two reasons. The first reason was to test the theory against the life expectancy data. It is one thing to draft a qualitative theory of a complex process such as changing life spans. It is quite another thing to get the qualitative theory to agree quantitatively with real-life data. This will not happen by chance. I wanted to know whether the theory would succeed with the biblical life expectancy data or be falsified by it.

The second reason, assuming success, was to be able to learn as much as possible about the nature of the MePA vitamin from these ancient data. For example, it was clear, as previously discussed, that MePA must have a long biological half-life. I wanted to move beyond this qualitative observation to an objective quantitative appraisal of the biological half-life. I wanted to get the best objective, quantitative estimate possible of this and other parameters from the available data.

A second component of this second reason was the desire to be able to follow some processes in greater detail. An example in this category is the "aging" process experienced by Noah. Noah lived 600 years in the Pre-Flood, then he lived through the Spike, and then he lived several centuries in the Decline. His natural daily dose of MePA went from high, to very high, to low. How did all of this affect his "agedness" along the way?

The biblical life expectancy data are the only data that presently exist on the effects of the MePA vitamin on human longevity. To exploit these data fully to the benefit of the planet's modern human population, a rigorous mathematical model was mandatory.

The Model

I constructed the mathematical computer model using Fortran. The source listing can be found in the appendix. The result of the model is shown in Figure 11. The model clearly succeeds in explaining the data. This demonstrates both that the theory is sound and that the biblical life span data are valid historical observations.

The sudden jump upwards in life expectancies between Adam and Noah may look like a mathematical glitch, but it is a real jump in life expectancies. It results from the Spike. People born immediately to the left of the jump have a normal Pre-Flood life expectancy of 926 years. These people die of "aging" just as the Spike is beginning. As a result, they just miss its beneficial effects. People born immediately to the right of the jump live long enough to make it into the Spike and benefit from it. In the Spike, their exposure to very high levels of MePA extends their life expectancies by several centuries.

The Eber–Peleg Drop is at the opposite end of this benefit to life expectancies caused by the Spike, as previously discussed.

Details

The model involves two components: environmental and physiological. The environmental component yields the concentration of MePA in drinking water. This is shown by the red line in the bottom graph in Figure 11. This is the driving function for the physiological component, which calculates the theoretical life expectancies shown as the solid blue line in the top graph.

For the sake of the model, the biblical longevity data were treated as point estimates of mean life expectancy at birth, as previously discussed.

The first seven data points were simply averaged, resulting in the height of the initial horizontal blue line. These seven points were oth-

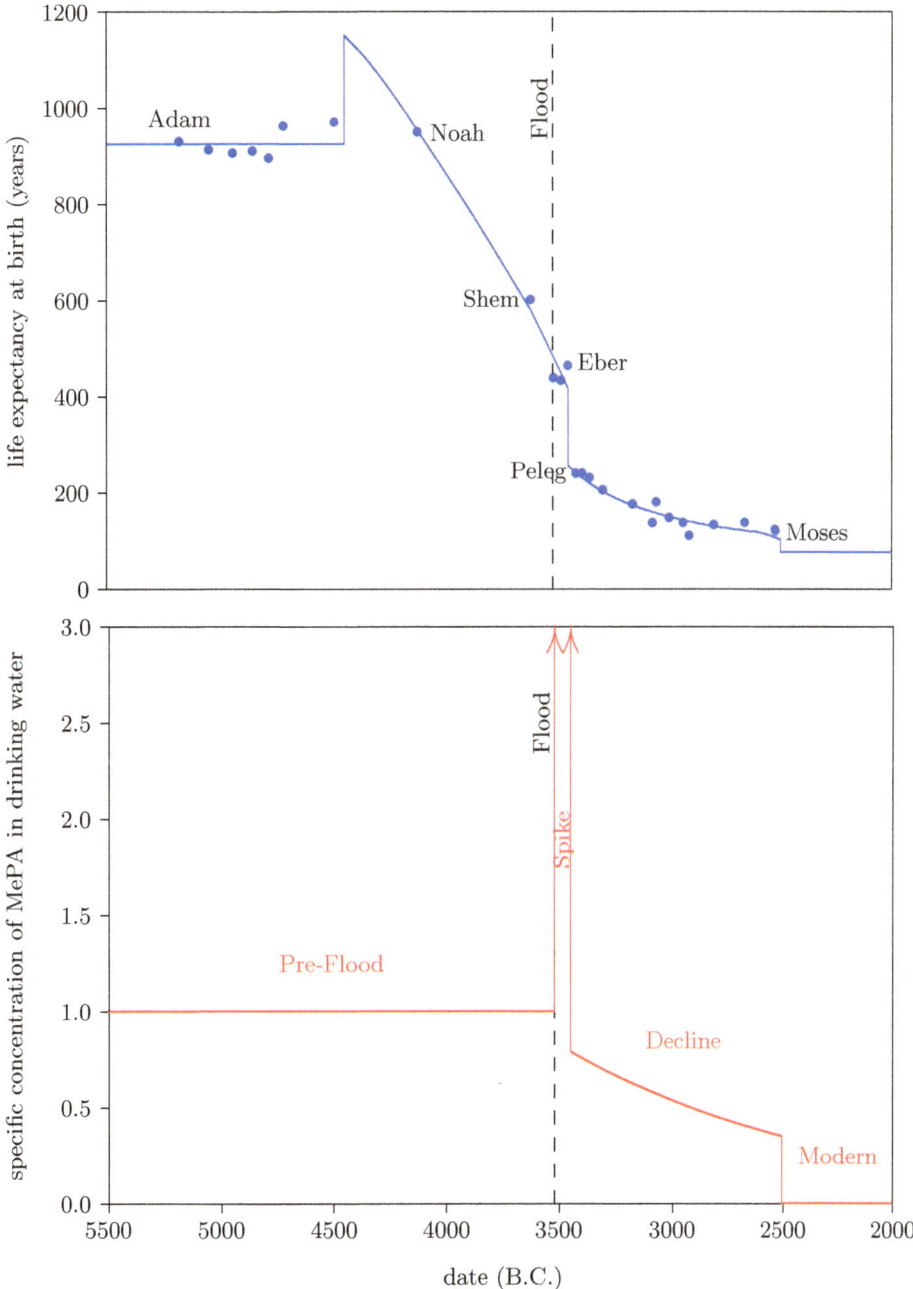

Figure 11: The results of the model are shown by the red and blue lines.

erwise excluded from the model. The line through the remaining data points results from a least squares fit of four free (i.e., adjustable) parameters. Two free parameters were needed to describe the changing environmental abundance of MePA, and another two were needed to describe the physiological response to varying doses of MePA.

Environmental Component

The environmental component of the model was divided into four distinct time periods, as previously discussed. The flux of MeP from the oceans to the atmosphere was assumed to be constant during the Pre-Flood, resulting in a constant supply of MePA in drinking water. All drinking water concentrations of MePA in all time periods were scaled with respect to this (unknown) Pre-Flood concentration, yielding the "specific concentration of MePA in drinking water" plotted in the bottom graph of Figure 11 (page 81).

The atmospheric concentration of MeP was very large during the Spike for reasons discussed above. Just how large is unknown, but we can get a rough idea as follows. Prior to the Flood, assuming steady state conditions, the rain of phosphorus-rich detritus onto the sea floor surrounding Antarctica supplied a flux of MeP to the atmosphere, resulting in unit specific concentration of MePA in drinking water. This MeP flux was halted as nearly the entire sedimentary supply of MeP was vented to the atmosphere by the Flood. The rain of phosphorus-rich detritus onto the ocean floor around Antarctica resumed after the Flood. At present, five and a half thousand years after the Flood, human longevity indicates that the specific concentration of MePA is still effectively zero. Subtracting 1000 years to allow for transport of MeP from the ocean floor to the atmosphere at present yields a minimum of 4500 years worth of Pre-Flood MeP vented to the atmosphere during the Flood. Thus, the flux of MeP to the atmosphere during the year of the Flood increased by at least a factor of 4500.

For such a large Spike, the actual amplitude turns out to be irrelevant. The physiological part of the model sets a fairly low cap on the body's reserve capacity of MePA, and this cap would have been met throughout this interval.

The end of the Spike interval coincides with the Eber–Peleg drop in life expectancies. This drop may have happened anywhere between the third and fourth data points following the Flood. The model gives a best fit when the drop occurs close to the third data point, with a total duration of 69 years for this interval.

The atmosphere was being provided with MeP from the oceanic reservoir during the Decline interval. The Decline interval ends with the Moses Drop in 2504 B.C., as previously calculated. This drop signals the exhaustion of the oceanic MeP reservoir.

The existence of the Moses Drop means that the oceanic supply of MeP came to an end suddenly rather than gradually. This implies that, in the (3520 - 2504 =) 1016 years it took Flood ocean bottom water to rise to the surface, this Flood ocean bottom water did not mix significantly with the underlying post-Flood ocean water. This, in turn, implies strong stratification between these two bodies of water, which implies that they had significantly different densities. Thus, the existence of the Moses Drop informs us that, immediately after the Flood, the density of Flood ocean bottom water was significantly less than the density of newly formed post-Flood ocean bottom water.

This is expected. It seems inevitable that the Flood would have effectively homogenized the Pre-Flood oceanic water mass. Normally, bottom water is more dense than any other ocean water because it is colder than any other ocean water. Surface waters are least dense, being warmed by sunlight. Homogenization of the Pre-Flood ocean by the Flood would have produced a body of Flood ocean water having a temperature and a density intermediate between normal bottom waters and normal surface waters. Thus the density of this well mixed body of waters from before the Flood would have differed significantly from the density of newly formed, post-Flood bottom water.

The idea that the Flood homogenized the oceans is supported by simple geometrical considerations. The average depth of the oceans is 3.8 kilometers. The average travel distance for ocean water during the Flood was roughly one quarter of earth's circumference during the waxing of the Flood, then another one quarter of a circumference back again at the waning of the Flood. This is a total distance of roughly 20,000 km. Thus the ratio of the depth to the distance traveled is less than one five-thousandth. It seems nearly impossible to maintain density stratification in such a geometry, and it seems especially so when one considers that the flow was over uneven ocean floor and continental terrain.

The two environmental free parameters mentioned above were needed to describe the specific concentration of MePA in drinking water during the Decline, which is modeled as a decaying exponential. Decay is due to internal oceanic chemical and/or biological processes consuming MeP precursor. The first free parameter sets the amplitude of the exponential, and the second free parameter sets its rate of decay.

In the Modern time period, the flux of MeP to the atmosphere is taken to be zero by the model. What we know about the nature of the Flood says that MeP must have been nearly emptied from its phosphorus-rich, anaerobic sedimentary source beds surrounding Antarctica. In addition, the global constancy of life expectancies due to "aging" since the Moses Drop strongly argues for this choice. For example, well water would be expected to be depleted of MePA due to sorption of MePA by soils as rainwater moved from the surface through the soil into the well, yet modern populations drinking water from rivers or lakes show no difference in life spans compared with populations drinking water from wells.

Physiological Component

The physiological component of the model is most easily constructed by analogy with a small gasoline engine. Wear of engine components results from inadequate lubrication. Wear corresponds to "aging" and the lubricant corresponds to vitamin MePA in this analogy. Lubricant is slowly lost, by being vaporized from the cylinder wall with each ignition cycle, for example. Lubricant is supplied to the machine at varying rates from an external source not subject to control by the machine. A low rate of supply of lubricant results in a deficient amount of lubricant in the machine's lubricant reservoir. A deficient amount of lubricant results in inferior lubrication and an increased rate of wear of parts. A high rate of supply of lubricant may fill the machine's lubricant reservoir. Once the reservoir is full, the machine can receive no more lubricant regardless of potential supply rate.

The lubricant reservoir corresponds to the body's reservoir for storage of MePA in this analogy. The size of this reservoir is the first free parameter of the physiological part of the computer model. The second free parameter is the lifetime of MePA in the body (corresponding to the lifetime of lubricant in the machine). Here again, exponential decay was assumed.

A new machine has no wear (i.e., agedness = 0). The machine runs (lives) until it reaches a critical level of wear (i.e., until agedness = 1), at which point it is worn out and no longer works (i.e., it dies).

The biblical longevity data provide two steady state regimes, allowing a linear relationship between rate of wear and lubricant level to be derived for use in the model. Letting SC be the specific concentration of MePA in drinking water, the first is the Pre-Flood period in which SC = 1 and the average life span for males is 926 years, and the second is the Modern period in which SC = 0 and the average life span for males is 76.8 years.[30]

This simple mechanical analogy to biological "aging" works well, but it is deficient in one respect. While self-repair (i.e., healing) is not possible mechanically, it is possible biologically. The model allows for this possibility by allowing the rate of wear to become negative for large enough lubricant levels (Figure 12).

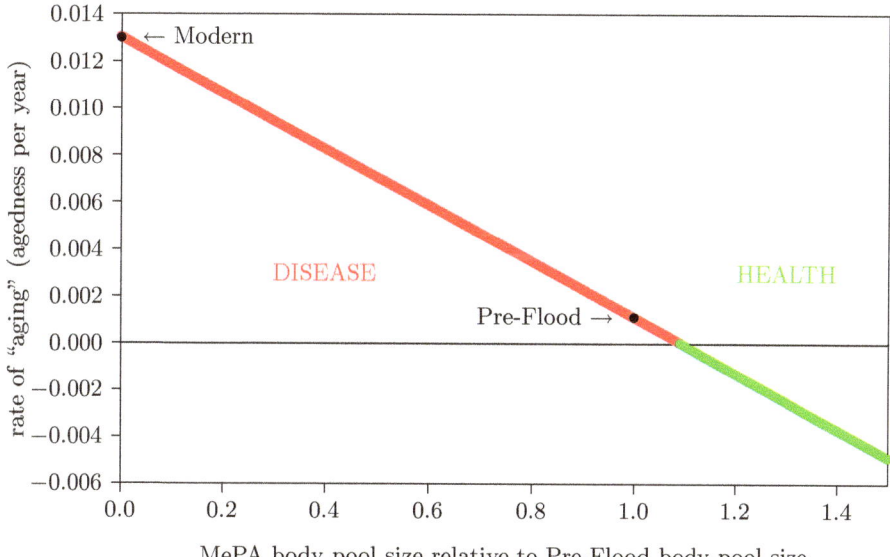

Figure 12: Rate of "aging" versus size of body pool of MePA. The red portion of the line results in increasing "agedness." The green part of the line results in decreasing "agedness" (i.e., healing of MePA deficiency disease).

---

[30]The average life span of modern males was calculated using data for U.S. males from the United States Social Security Administration's 2013 actuarial table, located at https://www.ssa.gov/oact/STATS/table4c6.html.

Results
Environmental Free Parameters
Duration of the Spike

The duration of the Spike was not a free parameter of the mathematical optimization of chi-square because it is a discrete, integer variable. Nonetheless, it could be optimized by giving it different values in separate runs of the computer model. When this was done, 69 years was found to be the duration yielding minimum chi-square, as mentioned above. This says that the large spike of MeP which the Flood released from the sediments surrounding Antarctica took 69 years to clear from the atmosphere.

Normally, MeP, which is easily oxidized, has a lifetime of only a few days at best in the atmosphere. The implication seems to be that the atmospheric chemical reactions which normally keep the atmosphere free of unwanted gases were significantly overloaded during the Spike interval. This overloading would have been contributed to by other anaerobic sea floor gases, especially methane.[31] Thus, it appears that earth's atmosphere may have been significantly polluted globally during much of the Spike.

Amplitude of the Oceanic Reservoir

The amplitude of the oceanic reservoir free parameter is the contribution of the oceanic reservoir to SC (the specific concentration of MePA in drinking water) in 3519 B.C., immediately following the year of the Flood. The model yields a value of 0.838±0.006 for this parameter. (All error estimates from the model are 1-sigma.)

If there had been no Flood, then the value of this parameter would have been 1.0, as it had been throughout the Pre-Flood interval. The fact that it is less than 1.0 implies that the oceanic flux of MeP to the atmosphere was reduced immediately following the Flood, compared to what it had been before the Flood.

Loss of MeP Within the Oceanic Reservoir

This parameter specifies the rate at which MeP was lost from the oceanic reservoir, not into the atmosphere, but internally due to chemical and

---

[31] John H. Seinfeld and Spyros N. Pandis, *Atmospheric Chemistry and Physics* (New York: John Wiley & Sons, Inc., 1998), Section 21.4.3, 1098–1100.

biological degradation. The model yields a value of $(8.63\pm0.07) \times 10^{-4}$ per year for this parameter. This means that it would take these internal processes about 900 years to deplete the oceanic MeP reservoir by a factor of two. Biological processes may have been dominant as "C-P lyase genes are abundant in marine microbes."[32]

Physiological Free Parameters
Size of the Body Reservoir

The model finds a reservoir size for MePA of 1.18±0.02 times the Pre-Flood body content of MePA (Figure 13). This says that the body reservoir for MePA was 15% underfull for Pre-Flood humans. To top up the body reservoir for MePA requires the dose of MePA to be 18% greater than the Pre-Flood dose, according to the model. It appears that this condition would have been easily met throughout the Spike interval, as discussed above.

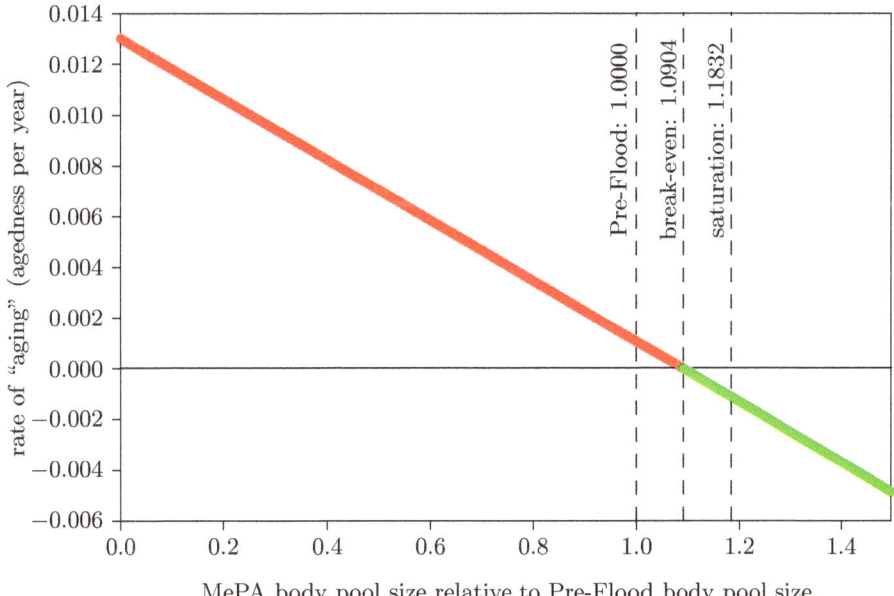

Figure 13: Rate of "aging" versus size of body pool of MePA. The Pre-Flood, break-even, and saturation body pool sizes are indicated by vertical dashed lines.

---

[32]William W. Metcalf et al., "Synthesis of Methylphosphonic Acid by Marine Microbes: A Source for Methane in the Aerobic Ocean," *Science* 337 (31 August 2012) 1104–1107.

Lifetime of MePA

The model finds a lifetime for MePA in the body of 194±4 years. The corresponding biological half-life is 135±3 years. This is a remarkably long biological half-life, which seems to imply specialized mechanisms for conservation of MePA and/or its metabolites.

Conclusion

It is clear, by visual inspection of Figure 14, that the fit of the model to the data is quite good. The model has succeeded in capturing the essence of the biblical life expectancy data and in quantifying its free parameters.

The success of the model drives home the conclusion that the biblical life span data cannot be mythological or otherwise fabricated. What did the ancient author of Genesis understand about methylphosphine (MeP), about its reactions with oxidants in the atmosphere yielding methylphosphonic acid (MePA), about Antarctica, about upwelling phosphate and nitrate in the southern oceans, about vitamins and vitamin deficiency diseases, about anaerobic digestion within sea floor sediments, about the baring of southern ocean floors during the Flood? Yet, all of these things and more must be properly understood to explain the pattern of these biblical life expectancy data points. The only reasonable explanation of the amenability of these ancient biblical data to quantitative, scientifically rigorous analysis is that they are accurately recorded, real life observations.

Conversely, the goodness of the fit to these data validates the model and its theoretical underpinnings. The central hypothesis is strongly corroborated: human "aging" is a vitamin deficiency disease resulting from dietary insufficiency of a previously unknown vitamin which was made globally deficient by the impact of Noah's Flood on earth's environment.

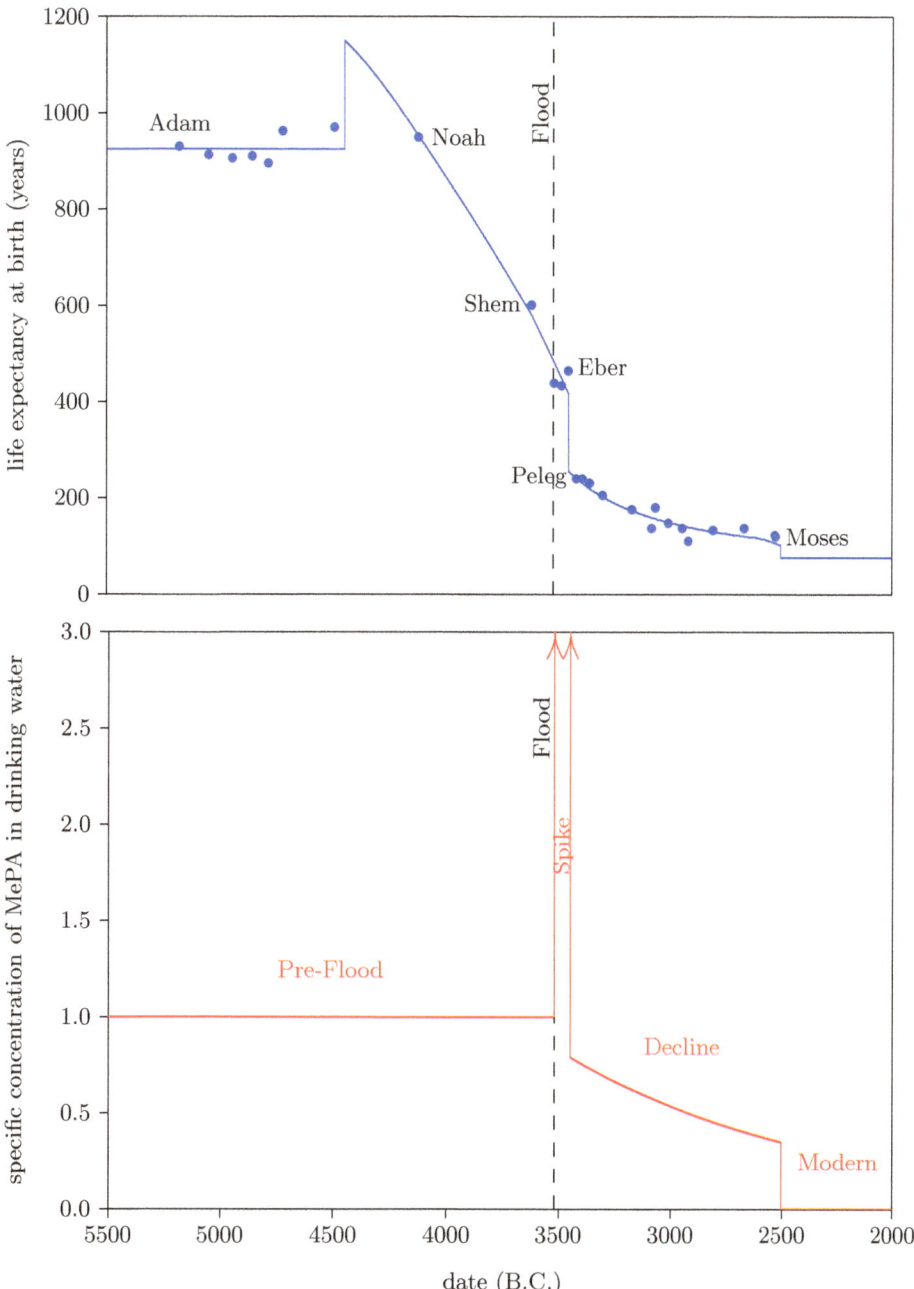

Figure 14: The results of the model are shown by the red and blue lines.

Chapter Eleven

Of Mice and Men

In the year 2000, when I was beginning to work on the human longevity problem full time, I wrote in my computer to-do list:

> Aging—the biblical aging data are everything. All effort should be geared toward understanding them and refining/correcting the mathematical model description of them. Experiment #1 is designed to see if gas X has sunk to such a low level that it no longer affects life span. This affects the proportionality between life span and gas X concentration in the atmosphere which the model assumes. We are pretty much obliged to assume mice are affected by gas X in this.

I did not know at the time that the process of "understanding the biblical aging data," and of "refining/correcting the mathematical model," would take 17 years. In fact, the model was rudimentary back then. I knew that there was an atmospheric gas involved, which I called "gas X." I did not yet understand that I was looking for a long lost vitamin—I was supposing that "gas X" was the whole target of the search. And though I had one thing right—that "the biblical aging data are everything"—I did not foresee just how correct that "everything" would turn out to be.

The Mice Versus Men Question

From the beginning of my research into the cause of human "aging," I had my doubts about whether mice and other small rodents could substitute for humans in laboratory experiments into "aging." There are, you have probably noticed, many differences between mice and men. Was it reasonable to suppose that human "aging" and mice "aging" were essentially the same thing?

My research eventually coalesced around the idea that human "aging" is a vitamin deficiency disease. This focused the mice versus men question to whether vitamin X should be expected to be efficacious with mice, dramatically lengthening their maximal life span the way it had dramatically lengthened maximal human life span prior to the Flood. The fact that not all animals need all of the human vitamins made it clear that the answer might very well be no. Only a few animal species need vitamin C in their diets, for example.

Mice are attractive as research subjects for the longevity research laboratory because they have a life span of only two to three years. This makes life span experiments feasible with mice.

Even more attractive, in terms of life span, are fruit flies. They have a maximal life span of only a few months. But they obviously raise even bigger doubts than mice about their suitability as proxies for humans.

I was never able to resolve the question of the suitability of substituting mice for men from any theoretical plane. From very early on, in simple pragmatism, I opted to use both rodents and fruit flies in my research, hoping for the best. As far as the experimental side of the research program was concerned, there was simply no alternative.

The Mice Versus Men Answer

Over the decades, I have spent many thousands of dollars and countless hours on rodent and fruit fly life span experiments. I have never been particularly lucky, and I was not "lucky" in this instance. Neither mice nor fruit flies ever showed any evidence of life-lengthening despite all of the many vitamin X candidates I tried with them. In each instance, it was further theoretical progress with the biblical life expectancy data which falsified these candidates. None, for example, until MePA, was ever able to explain successfully how it had been broken by the Flood.

My "luck" with MePA treatments of mice and fruit flies did not improve. I found that, while MePA showed no signs of hurting either mice or fruit flies in any way, even at very high doses, it did not appear to help them either. Their life spans showed no gains (Figures 15 and 16).

Only recently have I come to understand why this should be so. In hindsight, it is really very simple, involving just two fundamental principles.

The first principle is that vitamins do not act alone.

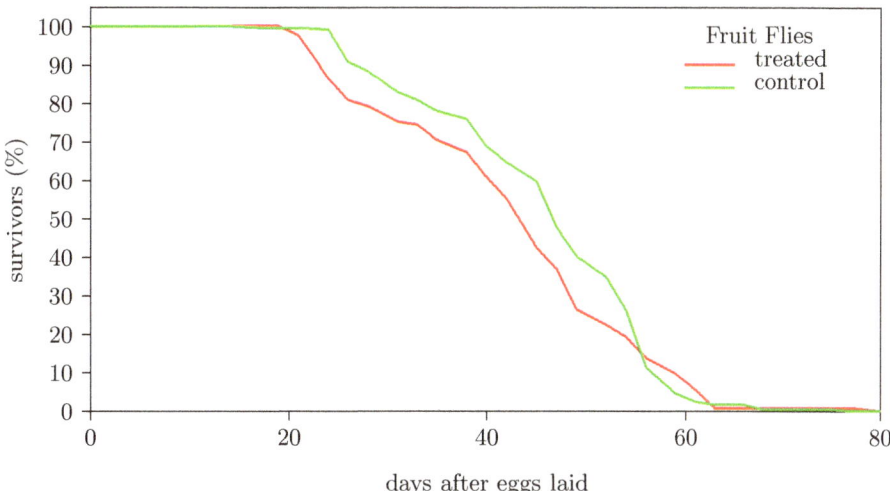

Figure 15: Survivorship curves of treated and control wild type fruit flies (Carolina Biological Supply). Flies were raised in colonies. Food vials were prepared using 3 grams of dried flake instant Drosophila media (Fisher Scientific) per 10 ml of 0.24% ethanol in distilled water for the control group, and 3 g MePA per liter of 0.24% ethanol in distilled water for the treated group. The control group contained 241 flies and the treated group contained 125 flies. The mean and standard deviation of the temperature for the experiment was 24.2±2.6 C.

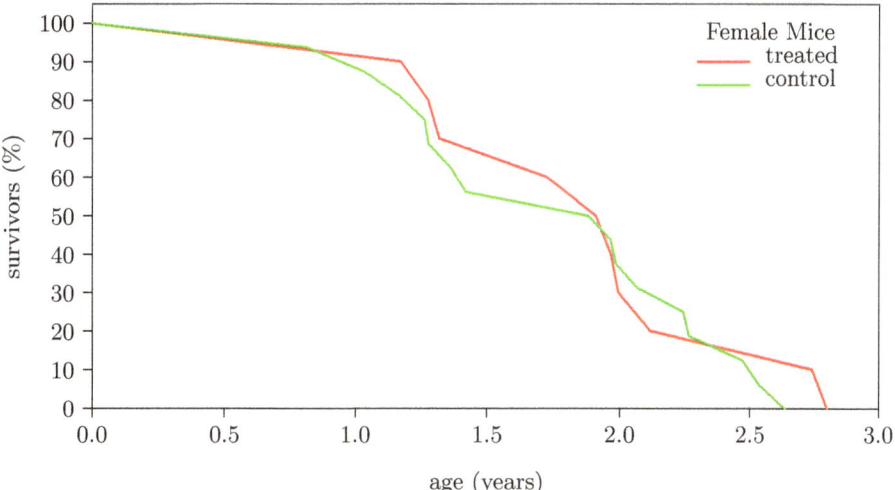

Figure 16: Survivorship curves of treated and control ICR (Harlan) female mice. The drinking water for treated mice contained 100 micrograms MePA/L. The control group contained 16 mice and the treated group contained 10 mice.

An automobile is a complex machine which allows you to move quickly from one place to another comfortably and safely. Without the key, however, the whole machine sits idle, doing nothing.

A vitamin is like an automobile key, setting in motion a biochemical machinery within a cell. Vitamins need the associated biochemical machinery to accomplish their specific tasks. They are not capable of operating alone any more than your automobile key is capable of moving you quickly from place to place all on its own without the automobile.

The second principle is "use it or lose it."

Eyesight has no practical utility inside dark caves. As a result, fish trapped in dark caves lose, over the course of many generations, the genetic coding which normally supports eyesight. Blind cave fish illustrate that loss of practical utility results in loss of the genetics necessary to implement that utility.

Putting these two principles together yields the answer to the mice versus men question.

MePA is the anti-aging vitamin. According to the biblical life expectancy data, it bestows on humans what today we call "superlongevity." Superlongevity has no practical utility for animals near the bottom of the food chain. In the wild, their life spans are determined by predation. Therefore, they will not possess the genetics necessary to implement superlongevity. To them, vitamin MePA is a key without an automobile.

Conclusion

I wish, of course, that I had been able to derive this result several decades ago, before I had spent all that time and money purchasing and caring for lab animals.

But this result is "unlucky" in other ways as well. It limits the demonstration of the life-lengthening efficacy of MePA to just the biblical/historical data we possess on humans. It means that there is no quick and easy modern "proof" of the theory. And it means that research into the biochemical processes underlying the MePA vitamin is deprived of the normal, short-life-span lab animals which dramatically aid research in so many other areas.

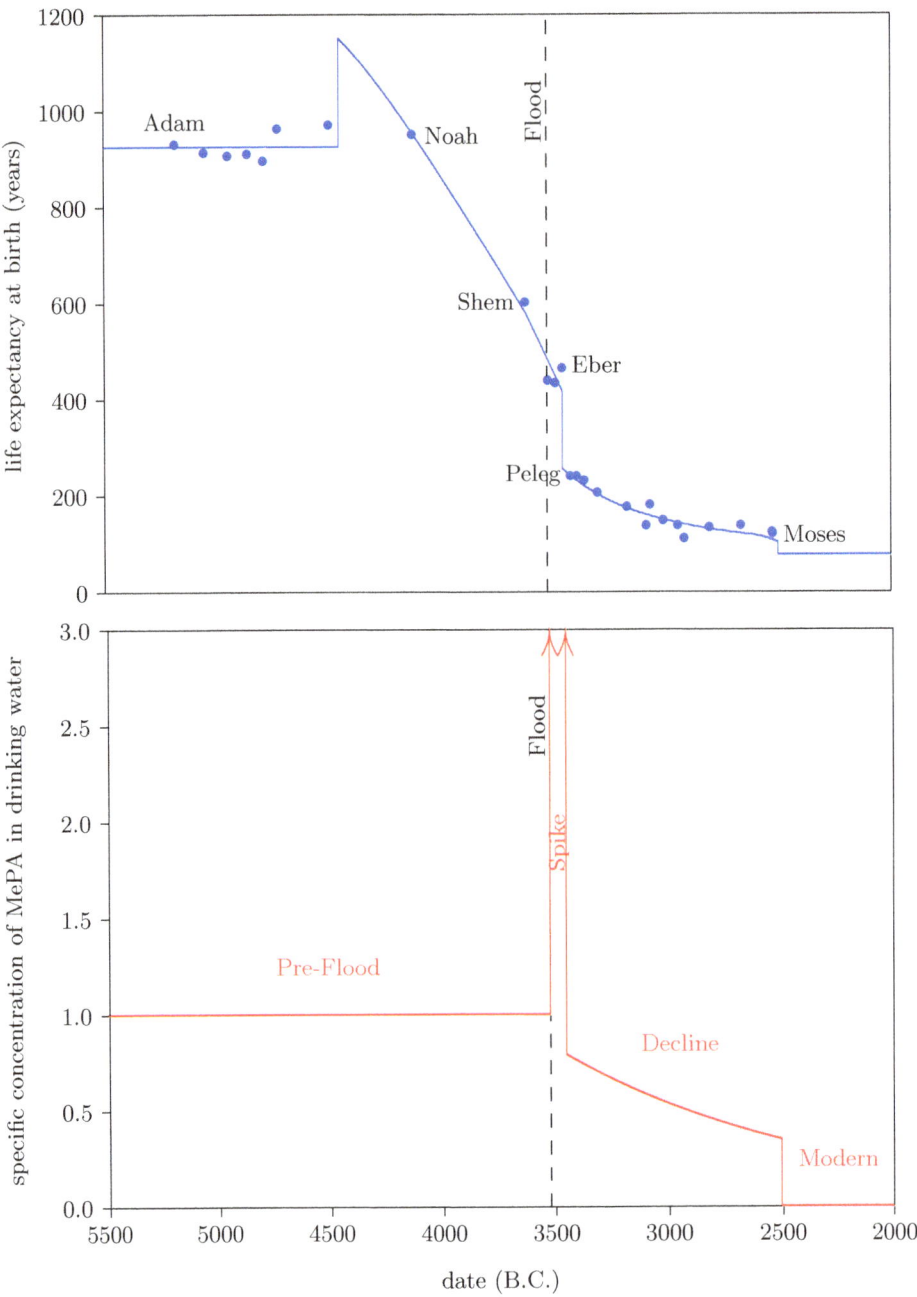

Figure 17: The results of the model are shown by the red and blue lines.

CHAPTER TWELVE

POTENTIAL LONGEVITY

For the youth will die at the age of one hundred, and the one who does not reach the age of one hundred shall be thought accursed. —Isaiah 65:20b.

The 926-year life expectancy at birth, enjoyed by Pre-Flood individuals, is an enormous improvement on our current 77-year life expectancy at birth. But the model says that it is possible to do better yet. The model says that Pre-Flood humans still suffered and ultimately died from MePA deficiency (i.e., from "aging"). Evidently, even though MePA was environmentally available before the Flood, its supply was still somewhat deficient. When this deficiency is corrected, as it was during the Spike, life spans exceeding a millennium become possible.

This first comes into view with the sudden jump in life expectancy from 926 up to 1151 years in 4444 B.C. (Figure 17). This jump is real, as mentioned previously. It results from the fact that individuals born close enough to the Spike lived into the Spike (assuming they did not get killed by the Flood), where they were subject to a very high MePA dose. This high dose topped up their body reservoir of MePA and thus increased their life spans.

The model says that the 926-year average life span of Pre-Flood humans was not a limiting life span for humans any more than our current 77-year average life span is a limiting life span for humans. There appears, in fact, to be no limiting life span for humans.

This is not to say that MePA grants immortality, for it most certainly does no such thing. MePA does not stop speeding bullets. MePA merely removes the particular cause of death we presently call "aging."

A Graphical Analysis

The increase in potential life span resulting from a proper dose of MePA in the diet is best illustrated graphically. Figure 18 shows a modern survivorship curve for United States males. The curve begins with a slightly sloped plateau. The plateau means that there are few deaths—almost everyone survives during the early years of life. The plateau persists, slowly increasing its negative slope, out to about age 55, followed by a noticeably accelerating rate of decline in the number of survivors. The ever-increasing rate of decline region is all due to MePA deficiency. People are dying of "old age" in this region. The probability of survival rapidly falls off. Only one person in 100,000 makes it to 111 years of age. The probability of anyone surviving beyond 125 years is essentially zero.

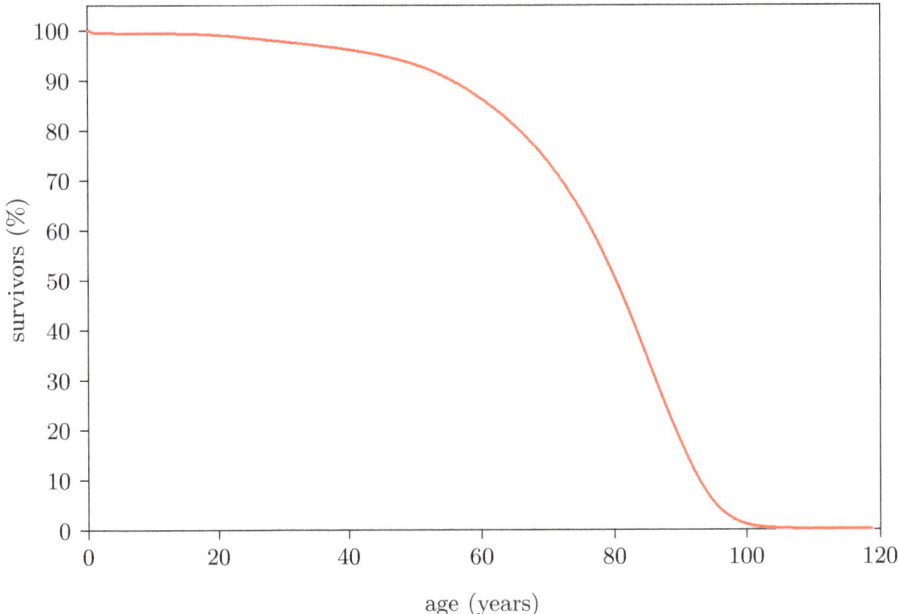

Figure 18: Survivorship curve for modern U.S. males. (Data for this graph are from the United States Social Security Administration's 2013 actuarial table, located at https://www.ssa.gov/oact/STATS/table4c6.html.)

When MePA deficiency is removed, the gradually sloping plateau region extends on into the future. Figure 19 shows the result when the present death rate at age nine pertains. For that case, only 1% fail to reach 100 years, and one in a thousand lives to 68,700 years. This hap-

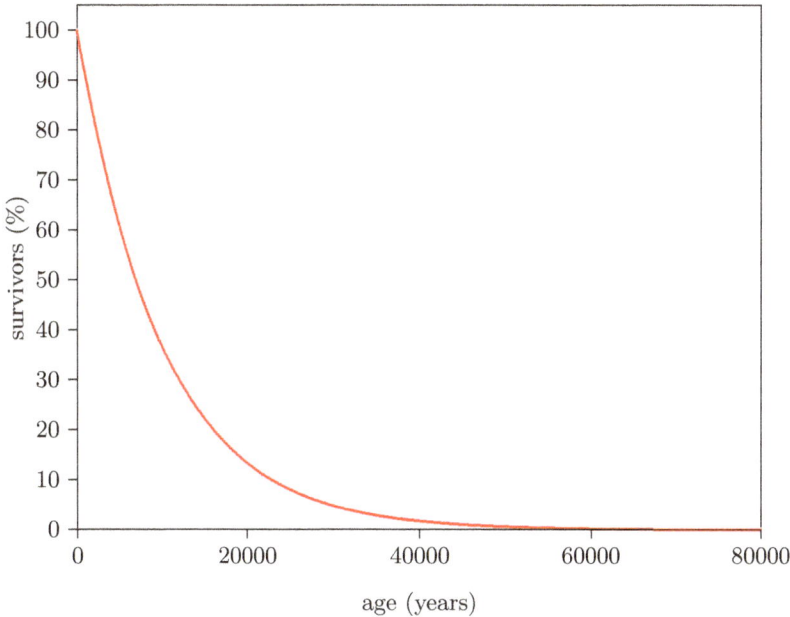

Figure 19: Survivorship curve for modern U.S. males once "aging" has been removed. Notice the change in the X axis scale compared to the previous graph.

pens because the probability of death no longer increases with chronological age. A 500 year old individual has the same probability of death in the next year as a 50 year old or a 5 year old. The multi-millenarian would not look any older than other mature individuals. Death in this case is due mainly to chance events such as car accidents, house fires, murders, lightning strikes, or any of a great number of other potential causes. According to this calculation, once MePA deficiency disease has been removed as the leading cause of death and replaced by the current death rate of nine-year-old U.S. males, a person will have about a 1 in 10 chance of living to 22,900 years of age, and a 1 in 100 chance of living to 45,800 years of age.

This calculation may fail in practice. The death rate I have chosen, one percent per hundred years, may be too optimistic. If hatred— ideological, political, personal— grows, its inevitable harvest of violence will no doubt escalate the death rate, ruining the calculation. But I am hopeful that warmer and wiser sentiments may begin to prevail in a world no longer dominated by *les enfants terribles*.

The calculation also assumes that the human body encounters no novel disease state out beyond 1000 years of age. We have no way of knowing what health challenges lie beyond 1000 years of age because nobody, ancient or modern, has been there that we know of. Who will be the first to cross this frontier into the unexplored reaches beyond?

Conclusion

Isaiah's prophecy, made some 2700 years ago, appears to be coming true in our lifetime. The future looks substantially brighter than the recent past for humans on planet earth. Instruct your broker to buy condominiums and sell crematoriums.

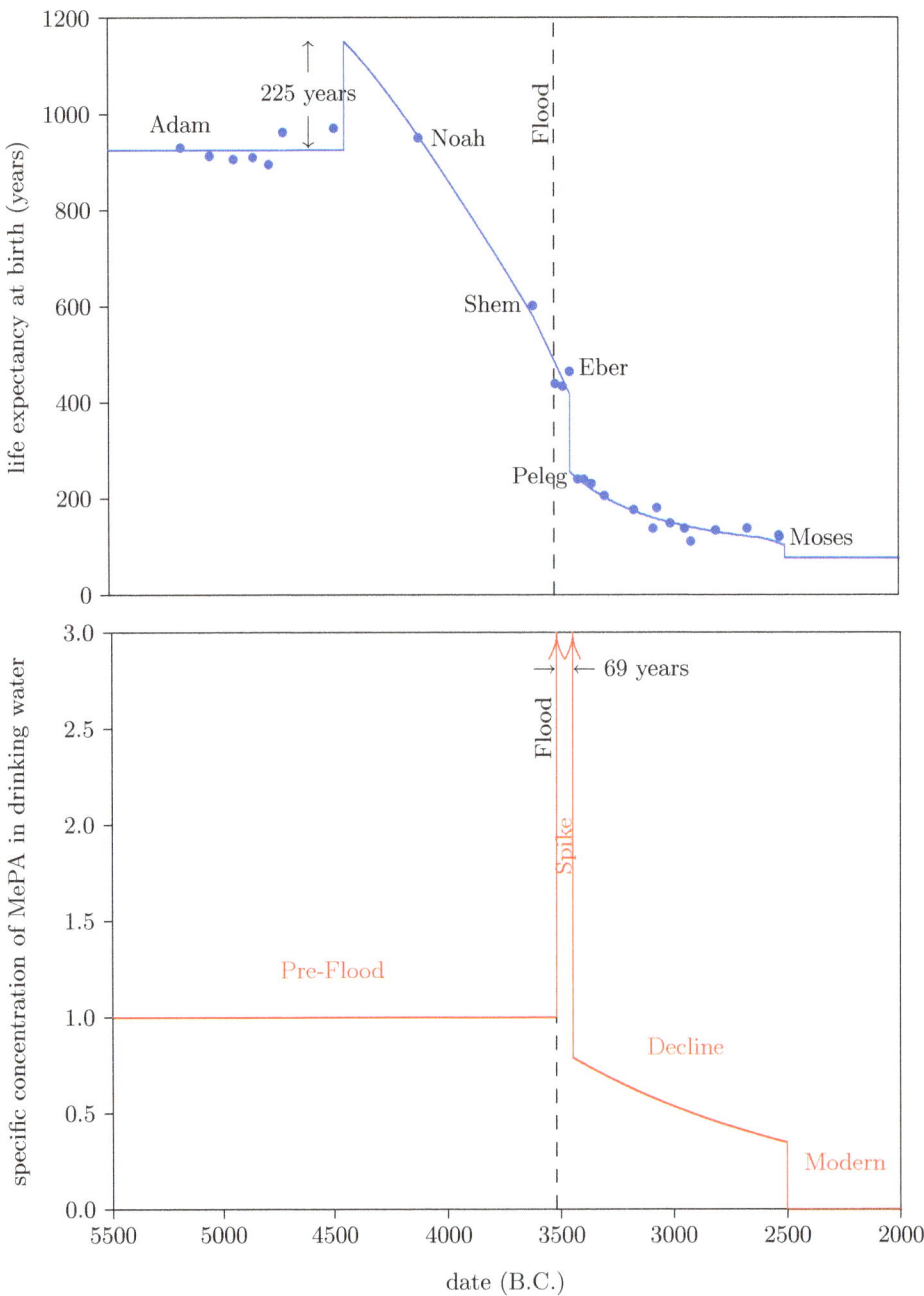

Figure 20: The results of the model are shown by the red and blue lines.

CHAPTER THIRTEEN

GROWING YOUTHFUL

The model shows that healing of agedness does indeed take place at high enough doses of MePA. Notice that the life expectancy extension enjoyed by individuals living into the Spike was longer than the duration of the Spike itself (Figure 20). According to the model, the life expectancy extension for someone born in 4444 B.C. was 225 years. The duration of the Spike was just 69 years. It thus appears that the high dose rate during the Spike did not merely halt further "aging" during the Spike. It appears that "aging" actually reversed during the Spike—a net healing of damaged tissues took place during the Spike. Those who lived through the Spike grew more youthful during their years in the Spike.

To be perfectly clear, I do not mean by this that individuals who lived through the Spike grew chronologically younger. They, of course, grew chronologically older. I also do not mean that they grew less mature. Those that went into the Spike as adults came out of the Spike as adults.

I mean that those who lived through the Spike grew less diseased, making them look more youthful (i.e., less aged). They came out of the Spike looking more youthful (e.g., fewer wrinkles) and feeling more youthful (e.g., better overall health) than they had looked and felt going into the Spike. Their physiological agedness had decreased.

Physiological Agedness

We are accustomed to equating physiological agedness and chronological age, but they are not the same thing. Chronological age measures how many years have passed since a person was born. Physiological agedness measures how advanced MePA deficiency disease is in a person. We are accustomed to equating these two at present because the two start out together at zero. Physiological agedness then begins to progress at a roughly equal rate in all individuals, making it possible to guess an adult's chronological age based upon the extent to which his MePA deficiency disease has progressed.

While it is now necessary to break the careless habit of equating these two, this habit provides a simple and convenient scale for measurements of physiological agedness. It is immediately clear what I mean if I say that a person looks to be 100 years old, for example. It is clear that I mean that physiological agedness, due to MePA deficiency disease, has come very nearly to a terminal condition. It is also clear what I mean when I say that an individual having a chronological age of 700 years has a physiological agedness of 30 years.

I will make use of this convenient physiological agedness scale below. It is a measure of the progress of MePA deficiency disease, conveniently calibrated by our everyday experience at present.

Noah's Physiological Agedness

Noah provides an example of a Pre-Flood individual who lived through the Spike. Figure 21 shows Noah's physiological agedness versus calendar year. The model finds that Noah's physiological agedness at the coming of the Flood was 50 years. (His chronological age at that point was 600 years.) His physiological agedness began to drop immediately upon entering the Spike—MePA deficiency disease began to heal. (For those of us who have passed our prime, this is obviously very good news.) Healing continued for some years following the Spike. Noah's body reservoir of MePA had been filled during the Spike and only slowly depleted of that filling following the Spike. By the time the healing benefit of the Spike had ended, Noah's physiological agedness had reduced to 42 years. Noah's body was made 8 years less physiologically aged because of having experienced the very high dose regime of the Spike.

Conclusion

While the fact of healing is very good news, the rate of healing is still sobering. Figure 21 shows that Noah's physiological agedness reduced just 5.6 years during the 69 years of the Spike. This says that the rate of healing is slow. It took Noah roughly 12 years of saturation dosing to reverse 1 year of physiological agedness.

It is clearly not a good idea to allow MePA deficiency disease to progress at all. This is most clearly seen by considering the case of the

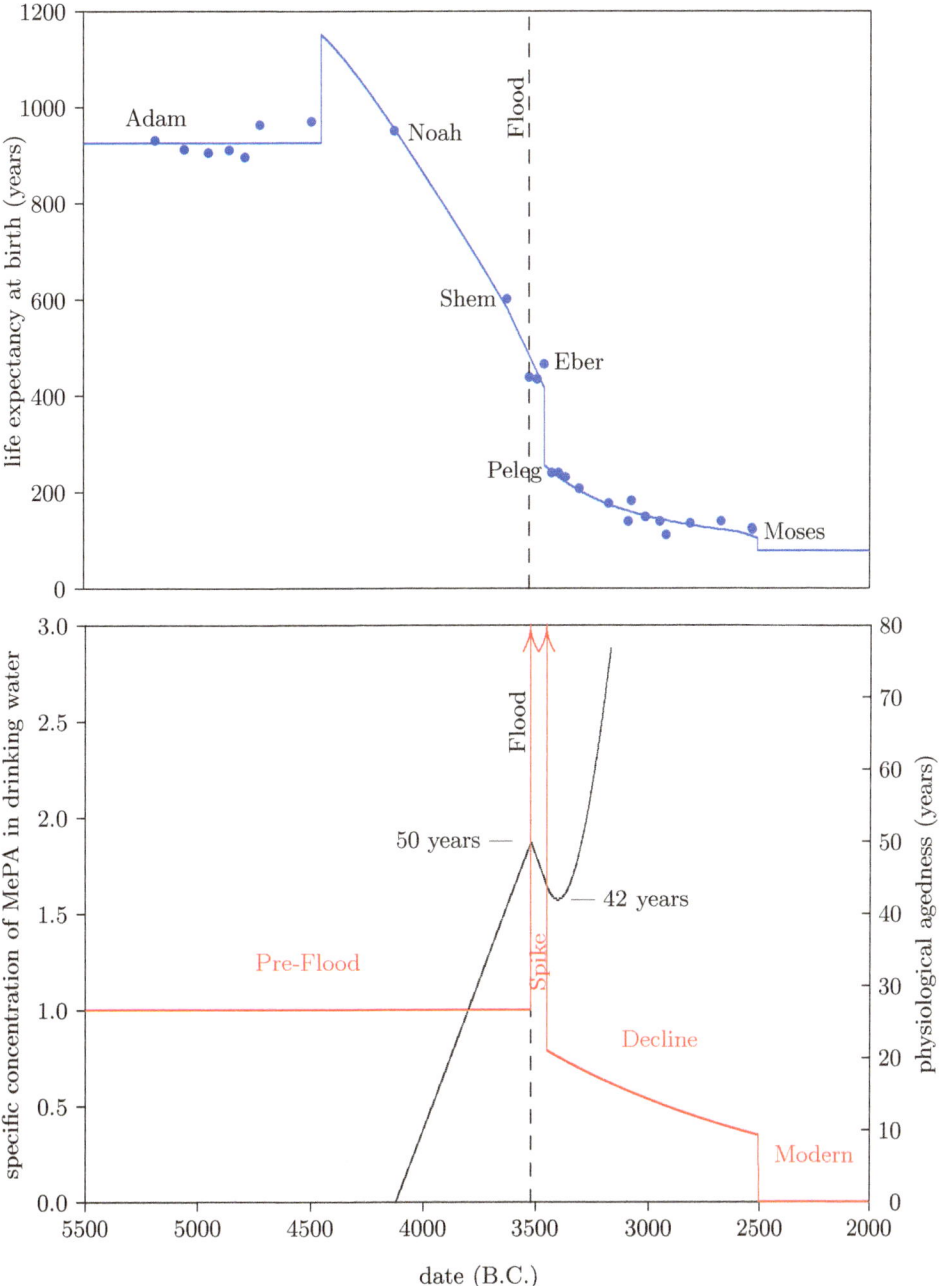

Figure 21: The solid black line in the bottom graph shows Noah's physiological agedness versus calendar year.

elderly. While proper dosing with MePA puts one immediately on the road to better health, an elderly person still has a significant probability of dying of complications arising from MePA deficiency disease lesions (of a heart attack, for example) while waiting for those lesions slowly to heal.

Chapter Fourteen

Dose Rate

Though the chemical compound corresponding to vitamin C, ascorbic acid, was discovered in the 1930's, the optimal dose is still an ongoing debate today.

> The appropriate daily intake is still vigorously disputed by scientists, and recommended allowances not only vary from one country to another; they also change from time to time within the same country. Although everyone agrees that the minimal daily requirement for vitamin C is 10 mg or slightly less, there is little agreement regarding recommended intakes.[33]

Determination of the optimal dose of any vitamin is not a trivial exercise. MePA is no exception. Determination of the optimal dose of MePA is likely to be an area of active research for years. At this early stage, only relatively crude guidance can be given.

Relative Versus Absolute Concentrations

The biblical life expectancy data are, once again, the best place to begin. But the modeling work discussed throughout this book was necessarily carried out using "specific concentrations." These are concentrations which are relative to the (presently unknown) Pre-Flood concentration of MePA in drinking water.

What is now needed is some means of converting these relative concentrations to absolute concentrations. We would like to know how many grams of MePA a person should take per day.

---

[33] "Scurvy and its prevention and control in major emergencies," *World Health Organization*, (1999) 17. http://www.unhcr.org/4cbef0599.pdf

The most direct route to converting the model's relative concentrations to absolute concentrations appears to be via the measurement of MePA concentration in Pre-Flood glacier ice. This has potential to tell us directly the dose rate prior to the Flood, which would allow replacement and saturation doses to be calculated by the model. Large efforts are presently being exerted to make these key measurements.

A second route to absolute dose rates is via the measurement of MePA in the urine and/or blood serum of treated human subjects. This method should allow the size of the body pool (or reservoir) of MePA to be measured, from which the necessary daily replacement dose can be calculated using the biological half-life provided by the model. Work in this direction is also presently underway.

Maximum Saturation Dose

Fortunately, the model provides a method of estimating the maximum possible Pre-Flood dose in absolute terms, and from this the maximum saturation dose can be calculated. This sets a useful upper limit for the recommended daily allowance.

The model gave a value of SC = 0.838±0.006 for the amplitude of the oceanic reservoir free parameter, as discussed previously. (Recall that SC is the specific concentration of MePA in drinking water.) This corresponds to the post-Flood oceans, immediately following the Flood. These oceans had been homogenized by the Flood, as previously discussed. As a result, they could not have contained any more than a surface saturation concentration of MeP. Any amount of MeP above this would have bubbled out of solution during homogenization.

The behavior of a gas in a liquid is governed by Henry's Law. This scientific law says that the concentration of the gas in the liquid will be proportional to the applied pressure of that gas. The proportionality constant for MeP does not seem to be readily available. I found 7.8×10^{-4} moles of MeP gas/(atm · cubic meter of water) when I measured this constant in my laboratory. This says that methylphosphine gas has a relatively low water solubility, similar to that of hydrogen gas.

Once the Henry's Law constant is known, the concentration of MeP in saturated surface water can be calculated. The result is 7.8×10^{-4} moles of MeP gas/cubic meter of water.

The average depth of the oceans is 3800 meters. The turnover time of the oceans after the Flood was the time from the Flood to the Moses Drop. This is (3520 - 2504 =) 1016 years. So the rate of turnover was 3.74 meters/year.

The surface area of the oceans is 3.61×10^{14} square meters. So MeP was being added to the atmosphere from the oceans at a rate of 1.0×10^{12} moles of MeP/year for SC = 0.838.

For the Pre-Flood, SC = 1. So the maximum Pre-Flood flux of MeP to the atmosphere was 1.3×10^{12} moles of MeP/year.

Now we need the atmospheric conversion efficiency for MeP to MePA. This is difficult to measure. I found a 23% conversion efficiency for reaction with ozone, and much smaller conversion efficiencies for both hydroxyl and nitrate radicals. However, my measurements were for the gas phase only, while some of the chemistry which converts MeP into MePA in the atmosphere probably takes place in aqueous phase on aerosols. This would likely boost the contribution of MePA from both the hydroxyl and the nitrate branches. Since we are presently trying to obtain a maximum possible Pre-Flood dose, it seems best to adopt a value of 50% as the probable maximum conversion efficiency. This yields a maximum of 6.3×10^{11} moles of MePA being produced in the atmosphere each year.

The total annual rainfall globally is 5.4×10^{17} liters/year. So the maximum Pre-Flood MePA concentration in rainwater was 1.2×10^{-6} mole MePA/liter, which is 1.1×10^{-4} grams MePA/liter. Assuming a consumption of 3 liters of water per person per day, and using the model's calculated relative saturation dose of 1.1832, yields a maximum saturation dose of 400 micrograms of MePA per day.

Vitamin Status Shown

This maximum saturation dose shows that MePA is indeed a vitamin. Vitamins are organic compounds, vital to the health of the organism, which are needed in tiny amounts only, and which must be obtained through the diet.

For the nine traditional water-soluble vitamins, recommended daily allowances range between 2.4 micrograms for vitamin B_{12} and 60 milligrams (i.e., 60,000 micrograms) for vitamin C.[34] Only three of the nine traditional water-soluble vitamins have a recommended daily allowance

---

[34] https://www.ncbi.nlm.nih.gov/books/NBK56068/table/summarytables.t2/.

at or below this 400 micrograms maximum saturation dose for MePA. These are folic acid (400 micrograms), biotin (30 micrograms), and B_{12} (2.4 micrograms).

Since 400 micrograms is a *maximum saturation dose*, not a recommended daily allowance for MePA, it is clear that MePA has a dose requirement which places it near the bottom of the recommended daily allowance range of the nine traditional water-soluble vitamins.

Vitamin X is truly a vitamin.

Ballpark Estimates

The maximum saturation dose calculated above should be regarded as a true maximum, not as a representative actual dose. It could easily be a factor of ten or more too high. The actual concentration of MeP in the oceans immediately post-Flood is not presently known. It may have been significantly below the saturation level the calculation assumed. In addition, the conversion efficiency of MeP to MePA in the atmosphere may be significantly below the 50% used in the calculation.

To help gain more of a feeling for what is likely to be a reasonable optimum daily dose of MePA, two additional methods of estimation are presented below.

Method 1: MSA

Measurements of methanesulfonic acid (MSA) in glacier ice may be used as a crude proxy for MePA.

MSA is a product of atmospheric chemistry acting on dimethyl sulfide (DMS), as mentioned previously.

DMS is emitted biogenically from the surface waters of the oceans. In contrast, MeP is emitted anaerobically from sea floor sediments, as discussed previously. The source strengths of these two processes can obviously be very different.

In addition, the efficiency for conversion of DMS to MSA in the atmosphere can obviously be very different from the conversion efficiency of MeP to MePA.

Nonetheless, the concentration of MSA in glacier ice provides a ballpark figure for what the biogenic ocean source plus atmospheric chemistry system is capable of.

Measured values from Antarctic cores show considerable variation geographically and temporally, but an average within a factor of two of 10 micrograms of MSA/liter seems representative.[35]

Assuming 3 liters of water consumed per person per day gives a ballpark (proxy) dose of 30 micrograms of MSA/day.

Method 2: Body Pool

A second ballpark estimate may be obtained using the body pool size of the traditional water soluble vitamins as an estimate of the body pool of MePA.

Loss of MePA from the body is assumed to be governed by the equation $X = X_0 e^{-\lambda t}$, where t measures elapsed time from termination of dosing. Thus $\delta X = -\lambda X \delta t$. This equation can be used to compute the daily replacement dose of MePA by taking X to be the normal body pool of MePA and δt to be 1 day.

The lifetime for MePA in the body has been calculated by the model to be 194 years. This gives lambda = 1/194 per year = 1.4×10^{-5} per day.

All that is needed to complete the calculation is the body pool size.

The largest body pool size I can find is for vitamin C at roughly 1500 milligrams.[36] The smallest is for vitamin B_{12} at roughly 2.5 milligrams.[37]

These two extremes for X give ballpark dose extremes of 20 and 0.033 micrograms/day respectively. This is a range of roughly 600, showing that they are indeed just ballpark estimates.

A more representative body pool size, based on vitamins B_1, B_6, and folic acid is roughly 25 milligrams. This yields a replacement dose for MePA of roughly 0.33 micrograms per day.

Conclusion

Taken together, these estimates suggest that the optimal dose of MePA will likely be found to be within a factor of ten of a few micrograms per

---

[35] See, for example, Saltzman et al., "Glacial/interglacial variations in methanesulfonate (MSA) in the Siple Dome ice core, West Antarctica," *Geophysical Research Letters* 33.11 (June 8, 2006).

[36] http://www.fao.org/docrep/004/Y2809E/y2809e0c.htm (accessed April 11, 2017).

[37] https://www.ncbi.nlm.nih.gov/books/NBK114302/ (accessed April 11, 2017).

day.

There seems to me to be less reason for concern about overdosing than there is for underdosing with MePA. Mice show no adverse effects, even at very high doses, as previously mentioned. For example, I have maintained one group of 36 ICR (Harlan) female mice at 100 milligrams MePA per liter in their drinking water for seven months now. This is equivalent to a human dose of roughly 250,000 micrograms per day. These mice are just over a year old. So far, there has been no mortality nor any sign of adverse health effects in any mouse of this group.

Water-soluble vitamins, especially those which are small acids, do not generally pose a risk of overdose, as any excess is readily removed from the body by the kidneys.

This property seems to be shared, as expected, by MePA. Tolerance to overdosing with MePA in humans seems to have been demonstrated by the full life spans enjoyed by both Noah and Seth who lived through the Spike, as well as by the first three post-Flood generations who were born during the Spike. All of these individuals appear to have been subject to very high MePA doses during the Spike, and these high doses seem to have done them significant good and no harm, at least as far as longevity is concerned.

Meanwhile, underdosing is known to produce the ravages of "old age," dramatically reducing potential life expectancy.

We tend to fear the unknown. The MePA vitamin is unknown relative to "aging," with which we have become comfortably familiar. But make no mistake, "aging" is the thing to be feared in this instance—it is the dread disease.

Prudence seems to me to call for immediate adoption of a 2 microgram MePA per day intake regimen for adults, pending recommended daily allowance refinement based upon further research.

Chapter Fifteen

Personal Testimonial

I began digging into the cause of pre-Flood longevity in my spare time in my early twenties, in the late 1970's. By 2000, I had laid sufficient groundwork to mandate that I make finding vitamin X my full-time occupation.

I first identified MePA as a vitamin X candidate, after many false starts, in September of 2014. I began computer modeling work on MePA in June of the following year. By late November 2015, tests with mice had revealed no signs of toxicity, and the theoretical case for MePA had become sufficiently strong to warrant testing it on myself.

It seemed clear that the risks of not beginning to take this compound were at least as high as any risks which might possibly be associated with taking it. I was reasonably sure by that time that MePA had been naturally present in the environment, and naturally present in everybody's diets for time out of mind prior to its loss in the Flood five and a half thousand years ago. This meant that I did not need to worry about it being like a new synthetic compound, totally foreign to the body, with potentially disastrous consequences.

The dose I would be taking—1 microgram/day—was miniscule. It seemed unlikely that many even synthetic compounds, chosen at random, would be able to do a body much harm at that low dose rate. To get a feeling for the low risk associated with such a small dose, compare, for example, to the deadly synthetic compound sarin, "used as a chemical weapon due to its extreme potency as a nerve agent."[38] A single dose of 500 micrograms (i.e., still tiny, but 500 times larger than the dose of MePA I would be taking) administered to a healthy male volunteer caused only "mild symptoms of intoxication."[39]

---

[38] https://en.wikipedia.org/wiki/Sarin (accessed July 5, 2017).
[39] https://en.wikipedia.org/wiki/Sarin#Diagnostic_tests (accessed July 5, 2017).

Meanwhile, I was approaching 61 years of age. My body was making it increasingly clear that I had passed the plateau phase of the human survivorship curve, and that I was rapidly on my way down its precipitous slope.

My own mortality had first begun to register with me when I had been forced to get prescription glasses early in my forties. I had always kept myself in reasonably good shape, without getting fanatical about it. I had never smoked, and I had stayed clear of alcohol. My wife and I had walked two miles or more together each day for decades, and I had found little trouble maintaining a healthy weight. But declining eyesight was just the beginning.

Subsequent years brought significant loss of hearing, hypothyroidism, worsening migraine headaches, and more. I had spent a great deal of time studying "old age." Now I was getting to experience first hand what I had studied.

The biggest age-related health problem struck just after I turned 50. It began with random pains. These were short in duration—one or two seconds—but intense, like being stabbed with a needle. I had these in a variety of locations, mostly extremities: thumb, big toe, etc. They seemed not to hit the same spot twice. I shrugged them off. Life was too short and its mysteries too absorbing to be distracted by petty annoyances.

Then, while raking gravel in the driveway, I found I was unable to maintain my grip on the rake handle. Peripheral weakness continued to spread over the following weeks, and then months. It would come and go, but each time it returned, it would be worse than before and stay longer.

After a couple of years of following false leads, I was finally diagnosed with CIDP (chronic inflammatory demyelinating polyneuropathy) by a talented neurologist. CIDP is a rare disease (a few per hundred thousand), most common in men over age 50.

CIDP is an autoimmune disease. The immune system attacks the myelin sheath surrounding nerves. The result is loss of nerve impulses to peripheral muscles, with ensuing weakness.

By the time I got the diagnosis, I was in very bad shape. I was having difficulty lifting my fork to feed myself, I could not button my shirt, I couldn't walk up or down stairs unassisted, and I was worrying at night about smothering under the blankets because I lacked the strength to move them off of my face.

While CIDP cannot be cured by modern medicine, its symptoms can generally be treated and controlled. Over the better part of the next decade, I went through several treatment regimens, from high prednisone to IV-Ig to Hizentra® home infusions, 60 ml twice per week. Relief of symptoms provided by treatment was not total, but it was nonetheless substantial, allowing me to live a somewhat normal life once again.

Eventually, I was able to resume walking two miles each day with my wife, Helen. But my leg muscles were no longer what they had been. They would rapidly grow weary and pretty much give out after the first mile. No amount of conditioning or exercise helped, so I began taking a bike with me. I would wheel it along beside me as I walked the first mile. It functioned as somewhat of a welcome "walker," keeping me more stable on my feet. Then I would ride it beside Helen as she walked the second mile.

I started taking MePA at 1 microgram/day on November 26, 2015. Three and a half weeks later, I was noticing positive health effects. Most significantly, I began to feel that I could do without the bike on our daily walks. I tried it and, sure enough, I could walk the entire two miles! CIDP had clearly begun to let up. This was eleven years after I had felt those first stabbing pains and subsequent peripheral weakness.

Another two weeks later, I was able to stop the biweekly infusions for CIDP that I had been obliged to be on for years. I had tried coming off the treatments in the past, only to be forced back on them if I wanted to stay out of a wheelchair. Had MePA healed my CIDP?

There were other possibilities. Spontaneous remittance is a characteristic of CIDP in the early years. But I was no longer in the early years. I had not experienced any spontaneous remittance since the relapse that nearly put me in a wheelchair prior to diagnosis. That had been nine years ago. One research study, aimed at assessing the long term prognosis of CIDP, concluded that prognosis "may be determined by the course and response to treatment in the first five years."[40] Based on the normal progression of this disease, remission seemed most unlikely. And the timing seemed just too coincidental. Remission had begun three and a half weeks following start of MePA dosing, after nothing but relentless

---

[40]S. Kuwabara et al., "Long term prognosis of chronic inflammatory demyelinating polyneuropathy: a five year follow up of 38 cases," *J. Neurol. Neurosurg. Psychiatry* 77.1 (Jan 2006): 66–70.

disease for years.

Another possibility was the well-known placebo effect. But this, too, seemed pretty unlikely to me. Following eleven years of disease, I had expected to have CIDP for the rest of my life. I had always regarded it as *possible* that vitamin X might reverse CIDP, but I had not regarded this as *probable*. CIDP is an age-related disease. Age-related diseases are results of "aging." It was clear from the start that, in the process of rolling back "aging," vitamin X might also roll back some age-related diseases. In my case, that meant that it might roll back CIDP. But through the years, when Helen and I had discussed the possible impact on CIDP of the yet-to-be-discovered vitamin X, I had expressed my sincerely held doubts that it could impact diseases of the immune system. I had started on MePA in this frame of mind. My purpose had been to halt "aging." Remission of CIDP had come as a bit of a (pleasant) surprise.

What was needed, to know for sure what was going on, was a proper clinical trial involving both treated and control groups. But all I had was myself. So we waited. Neither placebo effect nor spontaneous remittance would be expected to last more than a few weeks.

It has now been over a year and a half, and there has been no looking back. MePA does indeed appear to have cured my CIDP.

At the same time as CIDP was beginning to let up, I noticed that my skin was becoming more moist and supple. I'd had pretty severe eczema as a child, and my skin had tended to be dry and scaly into adulthood. The evident change for the better with my skin, together with the letting up of CIDP, got me really excited, because I knew that one of the symptoms of hyperthyroidism is oilier skin. I had been obliged to start taking levothyroxine for hypothyroidism early in my fifties. It seemed possible that MePA had begun to heal my thyroid, so that normal supplementation with levothyroxine was now causing hyperthyroidism.

I expected the next regularly scheduled blood test to come back showing high TSA. It came back normal and has remained normal. So MePA has made no discernable difference, so far, to my "worn-out" thyroid gland. Nonetheless, my skin has definitely improved.

I had always regarded a hot bath or shower as one of modern life's few affordable, worthwhile luxuries. I did some of my best thinking in the shower. But after age 50, I had found it increasingly necessary to stay out of the bath and out of the shower. The hot water would make my skin dry, rashy, and miserable the following day.

This too has now reversed. I am able to enjoy the luxury of a hot bath or shower pretty much as I please once again.

Once I had begun to take MePA, I also found that I was sleeping better. It is easiest to describe the change simply as a return to a more youthful sleep experience. Sleep was deeper and less interrupted. I needed less sleep, and I felt more rested.

Coming off of twice-per-week 60 ml infusions of Hizentra® complicates the question of what changes are due to MePA. For example, headaches have definitely decreased, but then headaches are a side effect of Hizentra®. For this reason, the experience of my wife, Helen, has been especially helpful.

Helen began taking 1 microgram MePA/day after I had been taking it for a year with no adverse side effects. She had lived close up with my CIDP for enough years to know that something real had happened. And she was concerned for her own future health, being aware from my research that MePA would not suddenly erase her own rapidly accumulating "old age" symptoms and debilitations.

Three weeks later she began reporting greatly improved sleep. Previously, she had been experiencing chronic sleep trouble. A typical morning had begun with, "I slept so terribly again last night, I don't know how I can keep up, feeling like this..." Now she was saying, "I haven't slept like this in years." Our grown children now comment on her remarkably improved ability to cope with life when they come home to visit. She replies, "Yes, what a difference a good night's sleep makes!"

Helen also corroborates the increased skin oils. In her case, the change is not so welcome as it has been for me. Her skin has always been pretty normal—definitely youthful compared to mine. The practical impact for her has been that she has had to begin washing her hair three times per week instead of two.

Another improvement is that we seem to heal more quickly. For example, I developed very sore muscles in my back between my shoulder blades, and opposite that on my front side, from a bunch of heavy lifting I had to do, on top of some unusual hand pulling of plants I had been obliged to tackle in the garden. The pain was of the sort which takes your breath away. I took nothing for the pain, and I did not treat the sore muscles in any way, which is pretty standard practice for me. I went to bed expecting to lose a lot of sleep due to pain whenever I rolled over. From years of previous experience, I expected at least three days of pain, followed by a week or two of slow healing of residual soreness. What happened is that I slept better than expected, and I woke up with only slight twinges of pain left. By that afternoon, even those were gone.

On the "no improvement" side, Helen has so far noticed no relief of rheumatoid arthritis symptoms. And neither of us has noticed any significant regression of gray hairs. But then, based on Noah's healing rate of just one physiological agedness year per dozen chronological years, we are not expecting to see rapid improvement in most things. We expect to be stuck in our physiological sixties for the next several decades.

But we *are* happy to have begun the journey back up out of this disease, and we *are* already feeling much improved. Our quality of life is simply better than it was for both of us before we began taking MePA. We now have to keep reminding ourselves not to overdo it—that we are still in our sixties.

Conclusion

The world is a very sick place at present. No doubt others have said this same thing in the past, but I am sure that none has ever meant it quite as literally as I do. The global population is sick and is dying of vitamin MePA deficiency disease. The cure for this sickness is now in hand. Human longevity may be restored globally by the simple expedient of inclusion in the human diet of micrograms per day of the long-lost, anti-aging vitamin, methylphosphonic acid.

APPENDIX

MeP_20170105.F95

The program listing below shows how I modeled the biblical life expectancy data based on vitamin MePA produced by atmospheric chemistry from MeP precursor gas generated in anaerobic sediments and delivered to the atmosphere via the oceans.

The code is in free-format Fortran 95. It was compiled as a 64-bit executable using GNU Fortran 6.1.0 via Simply Fortran by Approximatrix. The executable was run in a Windows 10 Command Prompt window on a PC having an Intel Pentium G3258 CPU.

```
    program MeP_20170105
!
! MeP_20170105 seeks to model the biblical life expectancy data based on the
!       precursor gas MeP (methylphosphine).
!   It uses a grid search to find a minimum of the chisqr hypersurface.
! It allows the width of the Spike interval to be easily adjusted, to optimize
!       this discrete integer parameter.
! The results are saved in the output file MeP.txt for plotting in the
!       spreadsheet MeP.xls.
!
! Units are everywhere mka (meters, kilograms, years).
!
    include "MeP_20170105.inc"
!
! Input biblical life span data; see MeP.xls for the source of these data.
! The life spans of the first 7 Pre-Flood patriarchs are averaged separately.
! They are not part of the data being fit by the program and they are not
!       included in the following data statements.
!
    DATA jb /-4120, -3617, -3517, -3482, -3452, -3418, -3388, -3356, -3297, -3167,
```

```
     -3081, -3067, -3007, -2943, -2916, -2806, -2668, -2530, -2527/
        DATA LEb /950, 600, 438, 433, 464, 239, 239, 230, 205, 175, 137, 180, 147, 137,
     110, 133, 137, 123, 120/
        i_pts = 19
!
!    Compute standard deviations of life expectancy at birth.
!        The mean and standard deviation from pre-Flood data points (excluding Noah)
!           is 926 +/- 28.9.
!        The mean and standard deviation from modern data for U.S. males
!           [https://www.ssa.gov/oact/STATS/table4c6.html] is 76.8 +/- 16.8.
!     Interpolate between these two points.
        do 5 i=1, i_pts
        sigma(i) = ( sigma_LE_0 - sigma_LE_now ) / ( LE_0 - LE_now ) * ( LEb(i) - LE_now
     ) + sigma_LE_now
5       continue
!
!    * * * FIRST, WORK OUT MePA IN DRINKING WATER * * *
!
!    Model SC(i) from before the Flood to the present by breaking the timescale into
!          four time intervals.
!      The first is the Pre-Flood interval.  The second is the Spike interval.  The third
!          is the Decline interval.  The fourth is the Modern interval.
!    _ _ _ PRE-FLOOD _ _ _
!    The Pre-Flood extends from the distant past to the Flood in 3520 B.C.
!       C_0 is the Pre-Flood concentration of MePA in drinking water.
!          It is assumed to be constant.
!          All MePA concentrations in drinking water are normalized with respect to
!          it, yielding the specific concentration:  SC(t) = C(t)/C_0
!             C(t) is the time-varying concentration, equal to C_0 Pre-Flood.
!
!    Set SC to 1 Pre-Flood.
        do 10 i = i_start_date, i_Flood_date
           SC(i) = 1.0D+0
10      continue
!
!    _ _ _ SPIKE _ _ _
!      The Spike is bounded by the Flood on the left and by the plummeting of life
!          spans between Eber (born -3452) and Peleg (born -3418) on the right.
```

```
!         The Flood happened 3520 B.C., so t_Flood = -3520.
!         Thus, the Spike is 3520 - ( 3452 + 3418 ) / 2 = 85 +/- 17 years long.
!     SC must exceed 1 at the start of the Spike.
!         This results from the southern oceanic floor depressurizing.
! This would have released MeP to the atmosphere from the phosphorus-rich ocean
!         floor sediments surrounding Antarctica.
!     SC_Spike is the value SC jumped up to following the Flood.
! Theory says it will be very large.
! The model is insensitive to this parameter, as long as it is large, because
!         of X_max.
! So it is treated as a constant rather than as a free parameter.
! Excess MeP was steadily removed from the polluted Spike atmosphere by conversion
!         of MeP to MePA.
! At the end of the Spike interval, excess MeP had been entirely removed from the
!         atmosphere.
! Thus, the Spike ends 3520 - 85 = 3435 B.C. (+i_delta_Spike)
!_
!_ if (t .gt. i_Flood_date .and. t .le. i_last_Spike_date) SC(t) = SC_Spike
!_
     do 11 i = i_Flood_date + 1, i_last_Spike_date + i_delta_Spike
         SC(i) = SC_Spike
11    continue
! _ _ _ DECLINE _ _ _
! Following the Spike, MeP continued to be sourced to the atmosphere from the
!         oceanic reservoir of MeP until the oceanic reservoir was exhausted.
! This defines the Decline time interval.
! It is bounded by the end of the Spike on the left and by a second, smaller
!         plummeting of life spans following the birth of Moses.
! Moses observed, in Psalm 90, that life spans had declined to 75 years
!         during his life span.
! Moses lived to 120 years.
! A person dying of old age at 75 when Moses was 120 would have been born when
!         Moses was 45 years old.
! Thus, this is the latest possible date at which MePA in drinking water had
!         dropped to zero.
! The year following the birth of Moses is the earliest MePA could have dropped
!         to zero.
! Take the midpoint (when Moses was 23) as the best estimate of when MePA
```

```
!          actually dropped to zero.
! Moses was born 2527 B.C., so this interval ends 2504 B.C.
! The inventory of MeP in the oceans would probably have been reduced by the
!          Flood relative to the Pre-Flood inventory.
! So SC would fall to some value less than 1 at the end of the Spike (and start
!          of the Decline).
!          SC_ocean is the initial value, immediately following the Flood. It is a
!          FREE PARAMETER of the model.
! The ocean acts as an elevator, carrying MeP from the deep ocean up
!          ultimately to the surface where it vents to the atmosphere.
! MeP is expected to have been lost (due, for example, to having been eaten by
!          microbes) on its way to the surface.
! Model this loss as a decaying exponential.
! lambda_ocean is the decay constant for MeP loss within the ocean. It is a
!          FREE PARAMETER of the model.
!_
!_ delta_t = t - i_Flood_date
!_ if (t .gt. i_last_Spike_date .and. t .le. i_Moses_Drop_date) SC(t) = SC_ocean *
!          exp(-lambda_ocean * delta_t)
!_
! _ _ _ MODERN _ _ _
! The Modern time interval goes from the end of the Decline to the present.
! SC(i) is taken to be zero during this interval.
!_
!_ if (t .gt. i_Moses_Drop_date) SC(t) = 0.0D+0
!_
      do 12 i = i_Moses_Drop_date + 1, i_end_date
         SC(i) = 0.0D+0
12    continue
!
! * * * NOW WORK OUT MePA IN THE BODY AND CALCULATE AGEDNESS AND LIFE EXPECTANCY
!          AT BIRTH * * *
!
! MePA dose is assumed to be proportional to its concentration in drinking water.
!    delta_X is the annual dose of MePA.
!        Thus delta_X = constant_2 * SC(j).
!    X is the total pool of MePA in the body.
!        It is unknown, so normalize it with respect to the pre-Flood pool, X_0.
```

```
!        This means that it starts out pre-Flood with a value of 1.
!
!     tau_X is the mean life of MePA in the body.  It is a FREE PARAMETER of
!        the model.
!
!     In the Pre-Flood period, SC(j) = 1.
!         So constant_2 = delta_X pre-Flood.
!     Meanwhile, X pre-Flood = 1.
!         So 1 = exp(-lambda_X) + delta_X, from the general equation
!         X = X * exp(-lambda_X) + delta_X.
!     Thus, delta_X = 1 - exp(-lambda_X).
!         emlx = exp(-lambda_X)
!         constant_2 = 1.0D+0 - emlx
!
!     Thus X = X_previous_year * exp(-lambda_X) + delta_X, where delta_t has been
!         taken to be 1 year.
!
!     X_max is the maximum size the MePA pool in the body can be.  It is a FREE
!         PARAMETER of the model.
!
!     Finally, X_birth, the size of the MePA pool at birth is needed.
!         This will be approximate---the baby will grow and hence deplete
!         the initial pool more rapidly than lambda_X would suggest.
!         I think the best simple approximation results from taking the initial
!         pool size proportional to the concentration in drinking water, SC(t).
!         This boils down to setting X_birth = SC.
!             For example, in Pre-Flood steady state, X_birth should clearly
!             be 1, and this corresponds to SC = 1.
!
! * * * NOW SEARCH FOR A MINIMUM REDUCED CHISQR * * *
!   Do this by a grid search of the hypersurface centered on the starting point.
!   Calculate step sizes for the grid search.
      lambda_ocean_start = lambda_ocean_0
      tau_X_start = tau_X_0
      SC_ocean_start = SC_ocean_0
      X_max_start = X_max_0
!
      degrees_of_freedom = i_pts - i_free
```

```
!
15     chisqr_keep = 1000.0D+0
!
       delta_lambda_ocean = lambda_ocean_start / sf
       delta_tau_X = tau_X_start / sf
       delta_SC_ocean = SC_ocean_start / sf
       delta_X_max = X_max_start / sf
!
       loops_total = ( 2 * i_steps + 1 ) ** i_free
       loops = 0.0D+0
!
!   Modify tau_X.
!
       do 50 i50 = -i_steps, i_steps
           tau_X = tau_X_start + i50 * delta_tau_X
           lambda_X = 1.0D+0 / tau_X
           emlx = exp(-lambda_X)
           constant_2 = 1.0D+0 - emlx
!
!   Modify lambda_ocean.
!
       do 40 i40 = -i_steps, i_steps
           lambda_ocean = lambda_ocean_start + i40 * delta_lambda_ocean
!
!   Modify SC_ocean.
!
       do 30 i30 = -i_steps, i_steps
           SC_ocean = SC_ocean_start + i30 * delta_SC_ocean
!
!   Modify X_max.
!
       do 20 i20 = -i_steps, i_steps
           X_max = X_max_start + i20 * delta_X_max
           write(*,fmt='(a, f8.2, f12.5,a)', advance='no') 'percent complete, chisqr
= ', loops/loops_total*100.0D+0, chisqr_keep, achar(13)
           chisqr_1 = chisqr()
           loops = loops + 1
           if ( chisqr_1 .ge.  chisqr_keep ) goto 20
```

```fortran
            chisqr_keep = chisqr_1
            tau_X_keep = tau_X
            lambda_ocean_keep = lambda_ocean
            SC_ocean_keep = SC_ocean
            X_max_keep = X_max
            i_tau_X_keep = i50
            i_lambda_ocean_keep = i40
            i_SC_ocean_keep = i30
            i_X_max_keep = i20
            itotal = abs(i20) + abs(i30) + abs(i40) + abs(i50)
!
20      continue
30      continue
40      continue
50      continue
!
!   Set all parameters to the minimum chisqr point found.
!
      tau_X = tau_X_keep
         lambda_X = 1.0D+0 / tau_X
         em1x = exp(-lambda_X)
         constant_2 = 1.0D+0 - em1x
      lambda_ocean = lambda_ocean_keep
      SC_ocean = SC_ocean_keep
      X_max = X_max_keep
      chisqr_keep = chisqr()
      write(*,fmt='(a, f8.2, f12.5)') 'percent complete, chisqr = ', loops/loops_total
     *  100.0D+0, chisqr_keep
!
!   Save results for plotting.
!
      open(unit=12, file="MeP.txt")
      do 60 i = -5200, -2000
          write(12,fmt='(i12,3e12.5)') i, SC(i), LE(i), PA(i)
60      continue
      close(12)
      if (i_steps .eq. 0) goto 70
!
```

```
!       Repeat if different point found.
!       Same point found is a necessary condition to avoid division by zero when
!          estimating standard deviations below.
!
      if ( itotal .ne. 0 ) then
          lambda_ocean_start = lambda_ocean_keep
          tau_X_start = tau_X_keep
          SC_ocean_start = SC_ocean_keep
          X_max_start = X_max_keep
          goto 15
      endif
!
!  Now estimate the standard deviations in the fitted values of the free parameters.
!
!     Estimate the standard deviation of lambda_ocean.
!
      lambda_ocean = lambda_ocean_keep + delta_lambda_ocean
      chisqr_plus = chisqr()
      lambda_ocean = lambda_ocean_keep - delta_lambda_ocean
      chisqr_minus = chisqr()
      lambda_ocean = lambda_ocean_keep
      sigma_lambda_ocean = delta_lambda_ocean * sqrt ( 2.0D+0 / (chisqr_plus + chisqr_minus
  - 2.0D+0 * chisqr_keep) )
      print *
      print *,'i, lambda_ocean, sigma, pe = ', i_lambda_ocean_keep, lambda_ocean_keep,
  '+/-', sigma_lambda_ocean, sigma_lambda_ocean/lambda_ocean_keep*100.0D+0
!
!     Estimate the standard deviation of SC_ocean.
!
      SC_ocean = SC_ocean_keep + delta_SC_ocean
      chisqr_plus = chisqr()
      SC_ocean = SC_ocean_keep - delta_SC_ocean
      chisqr_minus = chisqr()
      SC_ocean = SC_ocean_keep
      sigma_SC_ocean = delta_SC_ocean * sqrt ( 2.0D+0 / (chisqr_plus + chisqr_minus -
  2.0D+0 * chisqr_keep) )
      print *
      print *,'i, SC_ocean, sigma, pe = ', i_SC_ocean_keep, SC_ocean_keep, '+/-',
```

```
    sigma_SC_ocean, sigma_SC_ocean/SC_ocean_keep*100.0D+0
!
!     Estimate the standard deviation of tau_X.
!
      tau_X = tau_X_keep + delta_tau_X
         lambda_X = 1.0D+0 / tau_X
         emlx = exp(-lambda_X)
         constant_2 = 1.0D+0 - emlx
      chisqr_plus = chisqr()
      tau_X = tau_X_keep - delta_tau_X
         lambda_X = 1.0D+0 / tau_X
         emlx = exp(-lambda_X)
         constant_2 = 1.0D+0 - emlx
      chisqr_minus = chisqr()
      tau_X = tau_X_keep
         lambda_X = 1.0D+0 / tau_X
         emlx = exp(-lambda_X)
         constant_2 = 1.0D+0 - emlx
      sigma_tau_X = delta_tau_X * sqrt ( 2.0D+0 / (chisqr_plus + chisqr_minus - 2.0D+0
  * chisqr_keep) )
      print *
      print *,'i, tau_X, sigma, pe = ', i_tau_X_keep, tau_X_keep, '+/-', sigma_tau_X,
  sigma_tau_X/tau_X_keep*100.0D+0
!
!     Estimate the standard deviation of X_max.
!
      X_max = X_max_keep + delta_X_max
      chisqr_plus = chisqr()
      X_max = X_max_keep - delta_X_max
chisqr_minus = chisqr()
      X_max = X_max_keep
      sigma_X_max = delta_X_max * sqrt ( 2.0D+0 / (chisqr_plus + chisqr_minus - 2.0D+0
  * chisqr_keep) )
      print *
      print *,'i, X_max, sigma, pe = ', i_X_max_keep, X_max_keep, '+/-', sigma_X_max,
  sigma_X_max/X_max_keep*100.0D+0
!
!  All done.
```

```
!
!  Ring bell to signal completion of run.
70         write(*,*)'Ringing bell.  Use ctrl-c to end program.'
80         write(*,100,advance='no') achar(07)
100          format(a)
        call sleep(10)
        goto 80
    end
!
!**********************************************************
!  * * * FUNCTION CHISQR * * *
!
function chisqr()
    include "MeP_20170105.inc"
    PA = 0.0D+0
!_
!_ delta_t = t - (i_Flood_date + 1)
!_ if (t .gt. i_last_Spike_date .and.  t .le i_Moses_Drop_date) SC(t) = SC_ocean *
!       exp(-lambda_ocean * delta_t)
!_
    do 20 i = i_last_Spike_date + i_delta_Spike + 1, i_Moses_Drop_date
        delta_t = i - (i_Flood_date + 1)
        SC(i) = SC_ocean * exp(-lambda_ocean * delta_t)
20  continue
!
!  This is the main calculational loop for life expectancy
    do 30 i = i_start_date, i_end_date - 500
        j = i
!    Agedness at birth is zero.
        agedness = 0.0D+0
!    MePA body pool at birth.
        X = SC(j)
!  Time-changing annual dose of MePA.
25       delta_X = constant_2 * SC(j)
!  Time-changing body pool size.
        X = X * emlx + delta_X        ! delta_t = 1 year.
        if (X .gt.  X_max) X = X_max
!  Time-changing agedness.
```

```
            agedness = agedness + m * X + b
            if (agedness .lt.  0.0D+0) agedness = 0.0D+0
            j = j + 1
!   Save physiological agedness of chosen individual; birthdate of individual is
!         specified in if statement below
            if (i .eq. -4120) PA(j) = agedness * 76.8D+0 !  Noah = -4120, Shem = -3617
            if (agedness .lt. 1.0D+0) goto 25
            LE(i) = j - i
30       continue
!
!  Calculate the reduced chisqr
!
         chisqr = 0.0D+0
         do 40 i = 1, i_pts
!  yfit = LE(jb(i))
            chisqr = chisqr + ( ( LE(jb(i)) - LEb(i) ) / sigma(i) ) **2
40       continue
         chisqr = chisqr / degrees_of_freedom
         end
!
!********************************************************
!  This is MeP_20170105.inc
!
!  real*8 is double precision.
         implicit real*8 (a-h,l-z)
!
!  User-specified parameters.
!
!  Date parameters.
         parameter (i_start_date = -5200)
         parameter (i_Flood_date = -3520)
         parameter (i_last_Spike_date = -3435)
         parameter (i_Moses_Drop_date = -2504)
         parameter (i_end_date = -1500)
!
!  Grid parameters.
         parameter (i_steps = 5)            !  number of grid steps to search out to on either
!          side of the starting point; enter zero to do just the starting point
```

```
      parameter (sf = 1000.0D+0)      !  scale factor for grid search; divides the start
!         point value to give grid spacing for that free parameter
!
!  Model parameters.
      parameter (SC_Spike = 30.0D+0) !  NOT a free parameter
      parameter (i_delta_Spike = -16) !  NOT a free parameter; allows duration of Spike
!         to be adjusted; duration = 85 + i_delta_Spike
!
      parameter (i_free = 4)              !  number of free parameters in the model
!    starting point for grid search of the free parameter space (followed by
!         parameter uncertainties from grid search)
      parameter (lambda_ocean_0 = 8.6295194193469055D-4)    !  +/- 0.072523784375094215
      parameter (SC_ocean_0 = 0.838028242273622249D+0)      !  +/- 0.0061298609207415975
      parameter (tau_X_0 = 194.26218274932114D+0)           !  +/- 4.2427546379676411
      parameter (X_max_0 = 1.1831834963273076D+0)           !  +/- 0.018412002518696499
!
!  Life expectancy parameters.
!        The mean and standard deviation from pre-Flood data points (excluding Noah)
!        is 926 +/- 28.9.
      parameter (LE_0 = 926.0D+0)        !  average pre-Flood life expectancy at birth
      parameter (sigma_LE_0 = 28.9D+0)   !  uncertainty in the average pre-Flood life
!        expectancy at birth
!      The mean and standard deviation from modern data for U.S. males
!         [https://www.ssa.gov/oact/STATS/table4c6.html] is 76.8 +/- 16.8.
      parameter (LE_now = 76.8D+0)       !  average modern life expectancy at birth for
!         U.S. males
      parameter (sigma_LE_now = 16.8D+0) !  uncertainty in the average modern life
!         expectancy at birth
!
      parameter (b = 1.0D+0 / LE_now)    !  intercept for rate of aging line
      parameter (m = 1.0D+0 / LE_0 - b)  !  slope for rate of aging line
!
!  The blank common block is used to define a global variable storage area.
!  Program constants and variables (except integers, which are at the very bottom)
      common // constant_2
      common // degrees_of_freedom
      common // emlx
      common // lambda_X
```

```
      common // SC_ocean
      common // lambda_ocean
      common // tau_X
      common // X_max
!
!     Program arrays.  Integer arrays come last.
!     Real arrays
      common // LE                     ! fitted life expectancies
      real*8 LE(i_start_date:i_end_date)
      common // LEb                    ! biblical and modern life expectancy at birth
      real*8 LEb(1:19)
      common // PA                     ! physiological agedness
      real*8 PA(i_start_date:i_end_date)
      common // SC         ! specific concentration of MePA in the atmosphere
      real*8 SC(i_start_date:i_end_date)
      common // sigma !  the standard deviations of the biblical life expectancy data
      real*8 sigma(1:19)
!
!     Integer arrays
      common // jb                     ! jb is the date for biblical life span data
      integer jb(1:19)
!
!     Integer variables
      common // i_pts
```

INDEX

1 Kings 6:1 .... 13, 35
1000 years 13, 14
2 Kings 6:1–7 37
Aardsma, Gerald E. ... 13–15, 27, 30, 35, 36, 68
Aardsma, Helen 115–118
Aaron 29, 31–33, 59, 60
    date of birth 32
Abraham 25, 29, 31, 33, 52
acid, small 65, 66, 75, 112
Acts 7:4 31
Adam .. 17, 25, 27, 29, 31, 33–35, 48, 54, 55, 57, 58, 61, 80, 81, 89, 96, 102, 105
advantages, research 35
age at death data 32
agedness
    healing during Spike 104
    healing of 85, 103, 104, 118
    physiological 103, 104
    slow healing of 104
aging 19, 21, 24, 91, 130
    a disease 21
    growing up vs growing old ... 18
    meaning of the word 18
    mice versus men 91
    rate vs MePA body pool size   85
    reversed during the Spike ... 103
    slow rate of healing 104
"aging" 18, 22, 34, 36, 39, 41, 44, 49, 51, 67, 68, 79, 80, 84, 85, 87, 88, 91, 92, 97, 99, 103, 112, 116
aging theory
    evolutionary 39

supernatural 36
vapor canopy 38
Amram 29, 31–33
    date of birth 32
anaerobic
    byproduct gases 68
    digestion 68, 72, 88
    microorganisms 68
analogy
    automobile key 94
    gasoline engine 84
Antarctica . 68, 71, 72, 74, 82, 84, 86, 88, 111, 121
    sea floor sediments 71
anti-aging vitamin 47, 94, 118
Arpachshad 25, 29, 31, 33
atmospheric
    conversion efficiency, MeP to MePA 109
    pollution during the Spike ... 86
    trace gas, phosphorus 72
    trace gas 64–68, 72, 73
automobile key analogy 94
average life expectancy 32
average life span .. 19–21, 28, 37, 55, 56, 85, 97
    in particular geographical locations 20
    in primitive living conditions. 20
    pre-Flood 55

ballpark estimates of optimum daily dose, MePA 110
Bible/science 13, 15

biblical birthdates............ 30, 31
biblical chronological numbers.... 27
biblical chronology 14, 27, 30, 35, 36
    date for Noah's Flood....... 27
biblical life expectancies
    Adam to Moses.............. 33
biblical life expectancy data.. 34, 36,
    38, 39, 41, 49, 55, 79, 80, 88,
    92, 94, 107, 119, 131
    mythological................. 67
    not mythological......... 27, 88
biblical life expectancy graph. 34, 35
biblical life span data. 22, 27, 28, 32,
    35–38, 80, 88, 119, 131
    for males versus females..... 28
    mythological................. 67
    not mythological......... 27, 88
    table of...................... 29
biological half-life................ 50
    MePA....................... 79
    of MePA..................... 88
    of vitamin MePA............ 88
    of vitamin X...... 50, 52, 56, 60
biological pathway
    methylated phosphorus...... 73
biomethylation................... 74
    of phosphorus............ 73, 74
birthdate data 32
birthdates, biblical 30, 31
blind cave fish................... 94
blood serum, vitamin MePA in.. 108
body pool
    estimation of vitamin MePA
        dose................... 111
    size for vitamin MePA .. 87, 108
body reservoir size
    for MePA.................... 87
    for vitamin MePA 87
break-even body pool size
    of vitamin MePA............ 87
British navy, scurvy.............. 42

canopy theory of aging........... 38

carbon cycle 72
catabolism 74
    of phosphonates.......... 73, 74
catastrophe-occasioned disease ... 39
cause of the Decline.............. 69
cause of the Spike................ 69
causes of deficiency diseases...... 42
cave fish, blind................... 94
central hypothesis............ 41, 44
chronic inflammatory demyelinating
    polyneuropathy (CIDP) 114
CIDP (chronic inflammatory demyelinating polyneuropathy) 114–117
CIDP
    and vitamin MePA 116
    cause of 114
    placebo effect............... 116
    prognosis................... 115
    remission................... 115
    treatments 115
computer model............ 80, 119
concocted life span data... 27, 67, 88
Conquest of Canaan.............. 14
conversion efficiency, MeP to MePA 109
copy error....................... 13
Craig, P. J. 74

daily replacement dose of vitamin
    MePA 108
Dalldorf, Gilbert.............. 44–46
date of birth
    Aaron 32
    Amram..................... 32
    Joseph...................... 32
    Kohath..................... 32
    Levi 32
    Moses 32
date of the Flood............. 30, 35
decline in human longevity....... 30
decline phase................ 19, 20
Decline
    cause of 69

time period explained........ 64
time period 60, 61,
    64, 69, 79, 81, 83, 84, 89, 96,
    102, 105, 120–122
deficiency disease . 41, 42, 44, 45, 47,
    51, 85, 99, 103, 104, 106, 118
    causes of 42
    malady X̄................... 41
    old age 47
    vitamin MePA. 85, 99, 103, 104,
        106, 118
    vitamin 42, 44–46, 88, 92
deficiency
    physiological agedness 103
    vitamin.................... 109
density of Flood ocean 83
depth of sea floor sediments 68
Deuteronomy 34:7................ 29
di-methylsulfoniopropionate...... 66
dietary vitamin X 63, 66
difficulty of research problem..... 23
diffusion, steady state............ 69
dimethyl sulfide.......... 66, 73, 110
dimethylphosphine oxide 75
dimethylphosphine 74
dimethylphosphinic acid.......... 75
discovery of missing millennium .. 35
discovery process of vitamin MePA 14,
    17, 91, 113
disease
    catastrophe-occasioned 39
    vitamin deficiency 42, 44–46, 88,
        92
distribution of vitamin X 63–65
DMeP........................ 74–76
DMePiA 75, 76
DMePO......................... 75
DMS................ 66, 72, 73, 110
DMSP 66
dose range, water-soluble vitamins 109
dose rate...................... 107
Drop
    Eber–Peleg 48, 49, 55, 59, 80, 82

Moses 59, 83, 84, 109
dry skin and vitamin MePA 116, 117
duration of the Spike........ 86, 103

Eber–Peleg Drop in life spans 48, 49,
    55, 59, 80, 82
Eber 29, 31, 33, 48–58, 60, 61, 81, 89,
    96, 102, 105, 120
eczema and vitamin MePA...... 116
Eddy, Walter H.............. 44–46
efficiency of atmospheric conversion of
    MeP to MePA 109
Egypt.......................... 63
Enoch.......................... 28
Enosh 29, 31, 33
environmental
    abundance of vitamin X . 53–58,
        60, 61, 64, 71
    compartment......... 52, 53, 59
    component of the model.. 80, 82
    free parameters 83, 86
    half-life of vitamin X 52
evolutionary theory of aging...... 39
Exodus 6:16..................... 29
Exodus 6:18..................... 29
Exodus 6:20..................... 29
Exodus 7:7................... 31, 32
Exodus of Israel from Egypt...... 30
experimental dose of vitamin
    MePA............. 113, 117
eyewitness knowledge 14

fabricated life span data .. 27, 67, 88
Flood ocean, water density....... 83
Flood of Noah 14
Flood
    date of.................. 30, 35
    hemispherical............... 68
    homogenization of ocean water 83
    nature of 36, 67, 68, 84
    real historical event.......... 14
flux of MeP to atmosphere, Modern 84
Fortran.................... 80, 119

free parameters
    environmental 83
    physiological 84
fruit flies 17, 92, 93
    use in longevity research 92
fundamental hypothesis 21

gas, precursor . 67, 69, 71, 73, 75–77, 119
gases, anaerobic byproduct 68
gasoline engine analogy 84
Genesis 5:3 31
Genesis 5:5 .... 17, 29
Genesis 5:6 31
Genesis 5:8 29
Genesis 5:9 31
Genesis 5:11 29
Genesis 5:12 31
Genesis 5:14 29
Genesis 5:15 31
Genesis 5:17 29
Genesis 5:18, 21 31
Genesis 5:20 29
Genesis 5:24 28
Genesis 5:25, 28 31
Genesis 5:27 .... 17, 29
Genesis 5:31 28
Genesis 7:11 31
Genesis 9:29 .... 17, 29
Genesis 11:10–11 29
Genesis 11:10 31
Genesis 11:12–13 29
Genesis 11:12 31
Genesis 11:14–15 29
Genesis 11:14 31
Genesis 11:16–17 29
Genesis 11:16 31
Genesis 11:18–19 29
Genesis 11:18 31
Genesis 11:20–21 29
Genesis 11:20 31
Genesis 11:22, 24 31
Genesis 11:22–23 29

Genesis 11:24–25 28
Genesis 11:32; 12:4 31
Genesis 11:32 29
Genesis 16:16 31
Genesis 21:5 31
Genesis 25:7 29
Genesis 25:17 29
Genesis 25:26 31
Genesis 29 and 30 32
Genesis 35:28 29
Genesis 41:46; 45:6; 47:9 31
Genesis 47:28 29
Genesis 50:22, 26 29
Genesis life span data 22, 27, 52
Genesis longevity data 21
Genesis record of human life spans 20
Genesis record of the Flood 14
geographical
    locations, life span in 20
    uniformity of vitamin X 63
gerbils 17
glacier ice
    MePA concentrations in 108
    MSA concentrations in .. 66, 110
global population 24, 63, 118
goodness of fit, model to data 88, 89
graph
    biblical life expectancies.. 34, 35
    Noah's physiological agedness 105
    sea-surface nitrate 70
    sea-surface phosphate 70
gray hair and vitamin MePA 118
growing up vs growing old 18

Hain, M. P. 71
half-life, biological 50
half-life, environmental 52
healing and vitamin MePA 117
healing of agedness 85, 103, 104, 118
    during Spike 104
    rate of 104
    slow rate 104

healing of vitamin MePA deficiency disease 104
hemispherical Flood 68
Henry's Law 108
Hizentra 115, 117
homogenization of ocean water by Flood 83
hot water cooling, temperature versus time 37
human life spans . . 17, 18, 20, 24, 25, 28, 30, 36–38, 46, 47, 63, 79
    decline in 30
human longevity . . 17, 18, 25, 28, 30, 32, 39, 47, 72, 80, 82, 91
    decline in 30
human survivorship curve 114
human tests with vitamin MePA 113, 115
humans, limiting life span for 97
hydroxyl radical 66, 109
hypothesis, central 41, 44, 88
hypothyroidism and vitamin MePA 116

immortality and vitamin MePA . . 97
iodine cycle 72
Iraq 63
Isaac 29, 31, 33
Isaiah 65:20b 97
Isaiah's prophecy 100
Ishmael 29, 31, 33
Israel 63

J. R. R. Tolkien 13
Jacob 29–33
Jared 29, 31, 33
Jenkins, R. O. 74
John 2:1–11 37
Joseph 29, 31–33
    date of birth 32
jump in life expectancies 80, 97

Kane, Joseph W. 50
Kenan 29, 31, 33

Kohath 29, 31–33
    date of birth 32
Kuwabara, S. 115

Lamech 28
Leonard Hayflick 24
les enfants terribles 99
Levi 29, 31–33
    date of birth 32
life expectancies
    Adam to Moses 33
    jump in 80, 97
life expectancy at birth 32, 34, 48, 54, 55, 57, 58, 61, 80, 81, 89, 96, 97, 102, 105, 120, 130, 131
life expectancy data
    biblical 34, 36, 38, 39, 41, 49, 55, 79, 80, 88, 92, 94, 107, 119, 131
    modeling results 86
    modeling 79–88
    mythological 67
    not mythological 27, 88
life expectancy graph
    Adam to Moses 34, 35
life expectancy 30, 32
life span data 30
    biblical males vs females 28
    mythological 67
    not mythological 27, 88
life span
    alteration of 19, 28
    average pre-Flood 55
    average 19
    humans, limiting 97
    in particular geographical locations 20
    in primitive living conditions. 20
    maximum 20
    Noah 55
    statistics 24
life spans
    decline in human 30

    Eber–Peleg Drop in.. 48, 49, 55, 59, 80, 82
    exceeding a millennium 97
    Moses Drop in... 59, 83, 84, 109
    why they changed 46
lifetime for MePA in the body 88, 111
limiting life span for humans 97
longevity data..... 21, 28, 63, 80, 85
longevity research
    fruit flies 92
    mice....................... 92
longevity, potential............... 97
Luke 8:22–25.................... 37

Mahalalel................. 29, 31, 33
malady $\bar{X}$ .. 22–25, 30, 36, 41, 44, 45
    caused by Noah's Flood 36
    nutritional deficiency disease . 41
    research difficulty of 24
    similarities to scurvy......... 45
    symptoms 22
    the primary health problem.. 23
mathematical computer model... 80, 119
maturation, rate of............... 19
maximum human life span 19
maximum possible life span 20
maximum saturation dose
    vitamin MePA......... 108, 110
meaning of the word aging 18
MeP_20170105.F95.............. 119
MeP.. 74–77, 82–84, 86–88, 108–110, 119, 121, 122
    large atmospheric concentration of during Spike 82
    loss within oceanic reservoir . 86
MePA 75–77, 79–82, 84–89, 92–94, 96–99, 102–113, 115–123, 128, 131
    and CIDP 116
    and dry skin........... 116, 117
    and eczema................. 116
    and gray hair............... 118
    and healing................. 117
    and hypothyroidism 116
    and immortality............. 97
    and rheumatoid arthritis ... 118
    and sleep................... 117
    ballpark estimates of optimum daily dose.............. 110
    biological half-life........ 79, 88
    body pool size 87, 108
    body reservoir size........... 87
    break-even body pool size of . 87
    concentrations in glacier ice . 108
    daily replacement dose...... 108
    deficiency disease, healing of 104
    deficiency disease... 85, 99, 103, 104, 106, 118
    discovery process 14, 17, 91, 113
    dose, body pool estimation . 111
    dose, MSA estimation 110
    early experimental dose 113, 117
    first tests with humans. 113, 115
    in the blood serum 108
    in the urine................. 108
    lifetime in the body..... 88, 111
    maximum saturation dose.. 108, 110
    optimal dose........... 107, 111
    overdosing versus underdosing 112
    Pre-Flood body pool size 87
    recommended daily allowance 108, 110, 112
    risk of overdose............. 112
    saturation body pool size of . 87
    toxicity experiments........ 112
    vitamin status of 109
MePiA......................... 75
MePO 75
Metcalf, William W....... 75–77, 87
methanesulfonic acid 66, 110
Methuselah 17, 29, 31, 33
methyl addition to phosphorus ... 74
methyl bromide 73
methyl chloride 73

methyl iodide 72, 73
methylated phosphorus 75, 76
    biological pathway 73
    gases 73
    vitamin X candidates 75
methylphosphine oxide 75
methylphosphine 74, 77, 88, 108, 119
methylphosphinic acid 75
methylphosphonic acid ... 75, 88, 118
mice 75, 91–94, 112, 113
    use in longevity research 92
    vs men as research models ... 91
microorganisms, anaerobic 68
Middle East 63
millennium missing from traditional
    biblical chronology 35
missing millennium, discovery 35
missing thousand years 15
model
    computer 80, 119
    environmental component of 80, 82
    goodness of fit to data ... 88, 89
    mathematical 80, 119
    need of 79
    physiological component . 80, 84
    validated 88
modeling the life expectancy
    data 79–88
Modern flux of MeP to atmosphere 84
modern medicine 19, 20, 115
modern science 19, 24
Modern time period .. 60, 61, 69, 81, 84, 85, 89, 96, 102, 105, 120, 122
Moses Drop in life spans . 59, 83, 84, 109
Moses Drop
    date of 59
    density of Flood ocean 83
    sudden loss of MeP flux 83
Moses 28, 29, 31–34, 48, 49, 54, 57–61, 81, 89, 96, 102, 105, 121, 122
    date of birth 32
MSA 66, 110, 111
    concentrations in glacier
      ice 66, 110
    estimation of vitamin MePA
      dose 110
multi-centenarians 25
mystery
    of "aging" 68
    of human longevity 17, 18, 25, 36
    of the true nature of the Flood 68
mythological life span data 27, 67, 88

Nahor 28
natural
    distribution of vitamin X . 63–65
    entry of vitamin X into diet . 63, 66
    synthesis of vitamin X ... 63, 65
nature of
    sea floors 68
    the Flood 36, 67, 68, 84
nitrate 71, 88
    radical 66, 109
    sea-surface, graph 70
nitrogen cycle 72
Noah's Flood .. 13, 14, 27, 30, 35, 36, 44, 67, 68, 88
    biblical chronolgy date of 27
    cause of malady $\bar{X}$ 36
    radiocarbon date of 27
Noah's life span 55
    graph 54
Noah's physiological agedness ... 104
    graph 105
Noah 17, 20, 25, 29, 31, 33, 34, 48, 49, 52, 54–58, 61, 79–81, 89, 96, 102, 104, 105, 112, 120, 129, 130
Nobel Prize-winning physicist 15
normal life span 20
number one health problem 23

Numbers 33:39 29
nutritional deficiency disease .. 41, 47
    malady X̄ 41
    old age 47

oceans
    turnover time after Flood ... 109
    turnover time 64
"old age" ... 21–23, 44–47, 50–52, 59, 98, 117
old age
    false label 21
    nutritional deficiency disease . 47
    similarities to scurvy 45
optimal dose
    of MePA 107, 111
    of vitamin C 107
optimum daily dose
    ballpark estimates for MePA 110
outgassing of sea floor sediments . 69
overdose risk, vitamin MePA 112
oxygen cycle 72

Pandis, Spyros N. 66, 72, 86
past decline in human longevity .. 30
Peleg ... 25, 29, 31, 33, 48–61, 81, 89, 96, 102, 105, 120
personal testimonial 113
phosphate 71, 74, 88
    sea-surface, graph 70
phosphinate 75
phosphinates 76
phosphine 73
phosphonates 73–76
    catabolism of 73, 74
phosphorus
    atmospheric trace gas 72
    biomethylation of 73, 74
    cycle, atmospheric component 72
    cycle 72, 73
    gases, methylated 73
    methyl addition 74
    methylated 75, 76

    trace gas 72
physiological agedness 103, 104
    definition of 103
    Noah's 104
physiological component
    analogy, gasoline engine 84
    of the model 80, 84
physiological free parameters 84
placebo effect, CIDP 116
plateau phase 19, 23, 114
pollution of the atmosphere during the Spike 86
post-Peleg Decline 59
post-Peleg regime 55, 56
potential longevity 97
Pre-Flood body pool size
    of MePA 87
    of vitamin MePA 87
Pre-Flood time period 60, 61, 67, 79–83, 85–87, 97, 102, 104, 105, 107–109, 119, 120, 122, 123
precursor gas .. 67, 69, 71, 73, 75–77, 119
    of vitamin X 69, 71, 76
precursor, source of 68
primitive living conditions, life span in 20
principle
    "use it or lose it" 94
    vitamins do not act alone 92
properties of vitamin X 49
prophecy, Isaiah's 100
Psalm 90:10a 59
Psalm 90 .... 19, 28, 59, 121

radiocarbon 13, 27
    date for Noah's Flood 27
rate of maturation 19
recommended daily allowance of vitamin MePA 108, 110, 112
remission of CIDP 115
research advantages 35

results of modeling study......... 86
Reu...................... 29, 31, 33
rheumatoid arthritis and MePA . 118
risk of overdose, vitamin MePA . 112
Roels, Joris..................... 73

Saltzman...................... 111
sarin.......................... 113
saturation body pool size of vitamin MePA.................. 87
scurvy................... 42–47, 51
    British navy................. 42
    similarities to malady $\bar{X}$..... 45
    similarities to old age........ 45
sea floor sediments 68, 69, 71, 72, 74, 86, 88, 110, 119, 121
    Antarctica................... 71
    depth of..................... 68
    outgassing of................ 69
sea floor, southern................ 71
sea floors, nature of.............. 68
sea-surface nitrate graph......... 70
sea-surface phosphate graph...... 70
sediments, sea floor... 68, 69, 71, 72, 74, 86, 88, 110, 119, 121
Seinfeld, John H.......... 66, 72, 86
septic tank of planet............. 71
Serug................... 29, 31, 33
Seth 29, 31, 33, 112
Shelah.................. 29, 31, 33
Shem 29, 31, 33–35, 48, 52, 54, 56–58, 61, 81, 89, 96, 102, 105, 129
Sigman, D. M................... 71
sleep and vitamin MePA........ 117
small acid............ 65, 66, 75, 112
small molecule.................. 65
source of precursor............... 68
southern sea floor................ 71
"special" groups.................. 20
Spike time period........... 60, 61, 69, 79–82, 86, 87, 89, 96, 97, 102–105, 112, 119–122, 130
    duration................... 103

Spike
    aging reversed.............. 103
    atmospheric pollution during 86
    cause of..................... 69
    duration of.................. 86
    healing of agedness......... 104
    large atmospheric concentration of MeP................... 82
steady state diffusion............. 69
steady state .. 60, 69, 72, 82, 85, 123
Sternheim, Morton M............ 50
structure of vitamin C molecule.. 43
sulfur cycle..................... 72
superlongevity............... 24, 94
supernatural theory of aging..... 36
survivorship curve
    for United States males...... 98
    fruit flies................... 93
    human.................... 114
    mice....................... 93
    U.S. males............... 98, 99
    with aging.................. 98
    without aging............... 99
synthesis of vitamin X........ 63, 65

table of biblical life span data.... 29
temperature vs time, bowl of hot water.................... 37
Terah................ 29, 31, 33, 52
testimonial, personal............ 113
The Biblical Chronologist......... 14
the Exodus.................. 14, 32
    of Israel from Egypt......... 30
    real historical event.......... 14
theory of aging, supernatural..... 36
Tibet........................... 20
time period
    Decline 60, 61, 64, 69, 79, 81, 83, 84, 89, 96, 102, 105, 120–122
    Modern... 60, 61, 69, 81, 84, 85, 89, 96, 102, 105, 120, 122
    Pre-Flood.................. 60, 61, 67, 79–83, 85–87, 97, 102,

104, 105, 107–109, 119, 120, 122, 123
Spike 60, 61, 69, 79–82, 86, 87, 89, 96, 97, 102–105, 112, 119–122, 130
TMeP 74
TMePO 74, 75
toxicity experiments, vitamin MePA 112
trace gas 64–68, 72, 73
 and vitamin X 65
traditional biblical chronology 13, 15, 35
trimethylphosphine oxide 74
trimethylphosphine 74
Turkey 63
turnover time, oceans 64
 after Flood 109

University of California 24
University of Toronto 15
urine, vitamin MePA in 108
"use it or lose it" principle 94

van der Donk, Wilfred A. 75–77
vapor canopy theory of aging 38
Verstraete, Willy 73
vitamin C 42, 43, 45–47, 51, 92, 107, 109, 111
 molecule 43
 optimal dose debate 107
 structure 43
vitamin deficiency disease 42, 44–46, 88, 92
vitamin MePA
 and CIDP 116
 and dry skin 116, 117
 and eczema 116
 and gray hair 118
 and healing 117
 and hypothyroidism 116
 and immortality 97
 and rheumatoid arthritis ... 118
 and sleep 117

ballpark estimates of optimum daily dose 110
biological half-life 88
body pool size 108
body reservoir size 87
break-even body pool size of . 87
daily replacement dose 108
deficiency disease, healing of 104
deficiency disease ... 85, 99, 103, 104, 106, 118
discovery process 14, 17, 91, 113
dose, body pool estimation . 111
dose, MSA estimation 110
early experimental dose 113, 117
first tests with humans. 113, 115
in the blood serum 108
in the urine 108
lifetime in the body 88, 111
maximum saturation dose .. 110
optimal dose 107, 111
overdosing versus underdosing 112
Pre-Flood body pool size of .. 87
recommended daily allowance of 108, 110, 112
risk of overdose 112
saturation body pool size of . 87
toxicity experiments 112
vitamin status of MePA 109
vitamin X 44, 46, 47, 49–61, 63–69, 71–73, 75, 76, 92, 110, 113, 116
 as a trace gas 65
 as product of a trace gas 65
 biological half-life. 50, 52, 56, 60
 candidates, methylated phosphorus 75
 distribution of 63–65
 environmental abundance of 53–58, 60, 61, 64, 71
 environmental compartments 52, 53, 59
 environmental half-life 52
 geographical uniformity of ... 63

    in drinking water............ 66
    natural distribution of.... 63–65
    natural entry into diet ... 63, 66
    natural synthesis of...... 63, 65
    pre-Flood level.............. 46
    precursor gas......... 69, 71, 76
    properties................... 49
    small acid................... 65
    small molecule............... 65
    synthesis of.............. 63, 65
    water-soluble................ 65
vitamin
    anti-aging........... 47, 94, 118
    definition................... 109
    water-soluble....... 65, 109, 110
vitamins
    daily requirement....... 44, 107
    do not act alone principle.... 92
    not gases.................... 65

water-soluble vitamin... 65, 109, 110
    dose range.................. 109

Yu, Xiaomin..................... 76

Also by Gerald Aardsma

A New Approach to the Chronology of Biblical History from Abraham to Samuel

The Exodus Happened 2450 B.C.

Noah's Flood Happened 3520 B.C.

Available through Aardsma Research & Publishing
412 N Mulberry St.
Loda, IL 60948-9651.

Order online through www.BiblicalChronologist.org.